Designing Technical and Professional Communication

This concise and flexible core textbook integrates a design thinking approach, rhetorical strategies, and a global perspective to help students succeed as technical and professional communicators in today's multimodal, mobile, and global community.

Design thinking and good communication practices are rooted in empathy and human values. This integrated approach fosters students' ability to address the complex problems they will face in their careers, where they will collaborate with people who present diverse expertise, cultures, languages, and values. This book introduces the knowledge and skills as well as agile activities that help students communicate on projects within local and global communities. Parts 1 and 2 introduce the strategies for design thinking, audience analysis, communicating ethically, collaborating professionally, and managing projects to define problems and implement solutions. In Parts 3 and 4, students learn to compose content in text and visuals. They learn to structure and deliver content by choosing the right genre and selecting effectively from the communication options available in today's multimodal environment.

Designing Technical and Professional Communication serves as a flexible core textbook for courses in technical and professional communication.

An instructor's manual containing exercises, sample syllabus, and guidance for teaching in a variety of settings is available online at www.routledge.com/ 9780367549602.

Deborah C. Andrews is Professor Emerita of English and former director of the Center for Material Culture Studies at the University of Delaware. She has published several articles, book chapters, and textbooks on business, professional, and technical communication, especially in an international context. These include *Technical Communication in the Global Community* and *Management Communication: A Guide*. A researcher, consultant, and speaker, she is the former editor of *Business Communication Quarterly*. More recently, she has integrated her communications interests with research in material culture studies. She coordinated and edited an anthology, *Shopping: A Material Culture Perspective*, based on a colloquium series. Her latest research project, from which she has derived several articles and two book chapters, is a broad study of how the physical environment of 21st-century workplaces fosters or constrains collaborative communication.

Jason C.K. Tham is Assistant Professor of Technical Communication and Rhetoric at Texas Tech University, where he co-directs the User Experience (UX) Research Lab and UX student organization. He is author of *Design Thinking in Technical Communication* and co-author of *Collaborative Writing Playbook*. He conducts studies and teaches courses in the areas of UX research, information design, instructional design, and digital rhetoric. His scholarship has been published in journals such as *Technical Communication*, *Technical Communication Quarterly*, *Journal of Technical Writing and Communication*, *IEEE Transactions on Professional Communication*, *Communication Design Quarterly*, *Computers and Composition*, *Journal of Business and Technical Communication*, as well as some edited volumes.

Designing Technical and Professional Communication

Strategies for the Global Community

Deborah C. Andrews
Jason C.K. Tham

Routledge
Taylor & Francis Group

NEW YORK AND LONDON

First published 2022
by Routledge
605 Third Avenue, New York, NY 10158

and by Routledge
2 Park Square, Milton Park, Abingdon, Oxon, OX14 4RN

Routledge is an imprint of the Taylor & Francis Group, an informa business

© 2022 Deborah C. Andrews and Jason C.K. Tham

Portions of this book were previously published by: Pearson Education, Inc.

Library of Congress Cataloging-in-Publication Data
A catalog record for this title has been requested

ISBN: 978-0-367-55492-7 (hbk)
ISBN: 978-0-367-54960-2 (pbk)
ISBN: 978-1-003-09376-3 (ebk)

DOI: 10.4324/9781003093763

Typeset in Univers
by Apex CoVantage, LLC

Access the Support Material: www.routledge.com/9780367549602

Brief Contents

Expanded Contents

Acknowledgments

Writing this book was, like the process we write about in the book, a collaborative endeavor. We would like to thank publicly some of the many people who contributed to making the book happen. First, we thank the reviewers who guided our thinking as we integrated design and communication theories and practices. These include Michelle Eble, Pam Estes Brewer, Peter Huk, Derek Ross, Allison Hutchison, and Erin Frost. We'd also like to thank our editors at Routledge, Grant Schatzman and Brian Eschrich.

We are grateful to Ann Hill Duin, of the University of Minnesota, for bringing us together and setting up our collaboration. Thank you to Sonia H. Stephens of the University of Central Florida and Daniel P. Richards of Old Dominion University for the willingness to share excerpts from their published article on story mapping and sea level rise in *Communication Design Quarterly*. In addition, Carol Luttrell provided significant advice based on her years of corporate experience internationally and teaching at the University of Delaware. Through personal conversations about visual communication and his connections with design-related professionals, design studios, and visual identity and branding agencies in London, Bill Deering contributed significantly to the book's underlying concept and framework. For their reviews and recommendations, we thank Rebekah Smith, Akshata Balghare, Meghalee Das, Omonpee (O.W.) Petcoff, Jiaxin Zhang, and Kenyan Burnham. In addition, Meghalee, along with Wilson Knight and Liane Vásquez-Weber allowed us to use their communication products as models.

In the wider circles of collaboration, we thank our colleagues and students at Texas Tech University and the University of Delaware, along with many others we have gathered with at conferences around the world, for their ongoing inspiration and insights.

Part 1
Communicating in the Global Community

1 Communicating by Design

Wherever you work in the 21st-century global economy, you are going to work in an environment poised for change. And whatever the physical space you work in, that place will ground a larger digital space where you will collaborate on projects with shifting networks of people who present diverse skills, expertise, cultures, languages, values, attitudes, and perspectives. Communication ties it all together. In a setting that encourages agile and creative thinking, you will need strategies for communication that are adaptable, adjustable to ever-changing situations, creative, and effective.

This chapter introduces a process called *design thinking* that will help you develop such communication strategies. It begins by describing the phases of design thinking as they further enhance well-tested approaches to communication. As an open, recursive, and collaborative activity, design thinking fosters your ability to address complex and messy problems in uncertain environments. That's just what you're likely to face in your career. In addition, and importantly, design thinking and good communication practices are both rooted in empathy and human values. The next sections of the chapter preview the cultural and ethical dimensions of communication situations. The chapter ends with a scenario of design-thinking-enhanced communication in action: improving the accessibility of an airport. The framework of design thinking as integrated with strategies for communication underlies subsequent chapters throughout *Designing Technical and Professional Communication*.

DESIGN THINKING

If you like to write, you'll find in this book strategies for increasing the pleasure by taking on the perspective of a designer in a new landscape for communication. If you've never liked to write and put it off as long as possible, this perspective will give you a new approach to integrating communication practices into your technical projects. The concept of design thinking was popularized by the Hasso Plattner Institute of Design at Stanford University, also known as the d.school, whose founders also established IDEO, the commercial design consultancy. The approach is now widely used by professionals in design, architecture, engineering, and business. Thinking like a designer can enhance every aspect of communicating as a technical professional.

The process is commonly discussed in terms of five recursive phases: empathize, define, ideate, prototype, and test (Brown, 2020). They are shown in Figure 1.1.

DOI: 10.4324/9781003093763-2

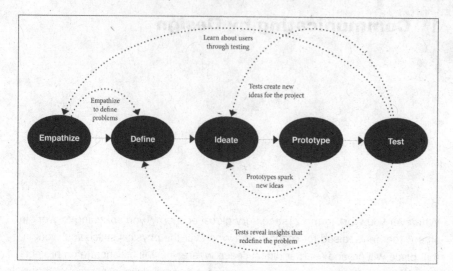

Empathizing

As a designer, you start by observing and listening to others, trying to understand what they are experiencing in their interactions with the environment. As a communicator, you can't know what to say, or to write, until you know *to whom* and *with whom*. Communicating in the workplace is functional communication. Within a context of empathy, of understanding other people's frames of reference, their perspectives, their experiences, such communication is meant to get work done. You communicate to achieve some purpose with and through other people.

You and Your Audience

These other people are readers of what you write. But reading these days increasingly means looking at pictures, viewing videos, listening to podcasts, sharing a multimodal environment. And communicating means more than writing. You compose documents and other products that also take advantage of multiple media and modes. The term now commonly used to designate the others you communicate with is your *audience*.

In composing, assess your own role as well as the role the audience plays in your shared situation. You may communicate as an individual or as a representative of one or many organizations. In a collaborative project, you may be both the source and the audience for various kinds of communication. You adjust what you say to each other as well as to the external audiences your collaborative endeavor addresses.

Often, you'll communicate in response to a direct request from the audience. The audience may ask you to answer a question, help them make a decision, or instruct them in the use of a system or tool. At the same time, you try to achieve your own purpose within that context. In response to a job listing, you write a cover letter and résumé that aim to persuade the reader to hire you. As a technical expert, you write a set of instructions to help customers use a new app you've developed. You help your audience to understand and to do. Try to imagine, or, better, actually experience, a setting in which your audience is using your communication product.

Design your approach to make that use as easy as possible. As an empathetic communicator, think about such questions as the following. You'll learn strategies for answering these and many other questions about your audience and communication situation in Chapter 2.

- *How skilled* is your audience at reading about the subject? In the role of a technical professional, you'll communicate with a variety of audiences, including those in your discipline or profession, sponsors and clients, investors and other stakeholders, community members, and the general public. Your audience will try to make sense and create meaning from what you say in light of their experiences, their values, their familiarity with terms and concepts, and their opinions about you and your organization. You will need to keep their level of expertise and interests in mind.
- *What else* do they know about the topic or situation you are addressing? Workplace communication begins in other texts, in numbers and images, and in conversations not aimed to produce a document but to solve a problem, to make new knowledge, to bring together multiple voices. Rarely are communication products one-time phenomena. New ones often reflect a series of documents on the topic, and you need to compose with that series in mind to control the range of possible interpretations of what you say.

Purpose

The audience will pay attention best if what they read or hear meets their purposes in reading and listening. Think about two broad categories of purpose: to explain and to persuade. To simplify, you write to *explain* when you know the audience needs to understand or to do something, perhaps because they have explicitly expressed that need. They'll pay attention because of that need. You structure your communication to respond to the need.

More difficult are situations in which an audience doesn't recognize a need that you see. You must first attract their attention and then *persuade* them to pay attention. You write to persuade when you want readers to act in a certain way, come to a decision you think is right, agree that something you think is significant is indeed significant, or approve a proposed course of action.

Defining the Problem or Opportunity

Your empathetic observation and understanding of others can help you initiate a project to make things better for them. At the same time, the project can make things better for you, for your organization, for an innovative new enterprise you envision, for your community, and for the environment, broadly understood.

Conventional Approach

In a routine or conventional approach to defining problems and getting work done, you think in terms of tasks. You have a set of skills, you know the rules, and you act in a well-defined way. The approach is highly structured. You've learned ways to solve such problems and apply them to a similar new problem. One researcher calls this approach to problem solving the application of "ideas you think *with*" (Hillier,

1996, p. 246, italics added). Such an approach has been significant for years in scientific and engineering research. It is systematic and ensures rigor and predictability. It still works.

Designing an Approach

Design thinking, on the other hand, is less focused on the routine and certain than on engaging creatively and collaboratively with messy and uncertain situations. Its goal is innovation. In the researcher's terms, it aims at creating "ideas you think *of*" (italics added). Emerging technologies often play a major role in the shift from conventional to design thinking. The technologies encourage you to see familiar situations in new ways. You think less in terms of tasks and more in terms of the resources you can engage with to create something better.

In crafting communications, you'll move between conventional and more customized or creative, design-enhanced approaches to suit a wide variety of situations. Your understanding of the audience will help you choose the appropriate response.

Conventions

Many scientific and technical documents reflect strict conventions of presentation. The individual or organization requesting the work—for example, a professor, a client, or a social agency—makes the rules. You follow. Learn and adhere to any conventions or standards that apply to what you are writing. These conventions ease both writing, as a fill-in-the-blank exercise, and reading. The reader knows where to find what's needed.

Genres

Match your audience's expectations, too, in the genre or type of document you are writing. Genres are the patterns that emerge within organizations or disciplines for writing in repeated situations. An approach to documenting original scientific research in a report or article, for example, is so well ingrained that it is known by its acronym, IMRAD: Introduction, Materials and method, Results, And Discussion. Other common genres are reports, proposals, business plans, manuals or instructional guides, white papers, briefings, and correspondence, including emails, messaging, and letters. Such common genres are discussed in detail in Part 4 of *Designing Technical and Professional Communication*. You'll use such genres extensively to arrange and present your work. Given the context of the global economy, while some genres appropriate in the US and Europe may be shared more broadly, others are not, and you should be alert to those differences.

For more about IMRAD reports, see Chapter 16

Designing in New Forms and Formats

In response to new technologies for delivering information to your audience, and new needs for information globally, you'll find yourself increasingly challenged to create new forms and formats. These communications may stretch the meaning and look of conventional genres or go off in a new direction, such as the "story maps" you'll read about in Chapter 8. Emerging from a network of writers, now often called "content producers," as well as designers and entrepreneurs, worldwide, these new genres reflect the new ideas for communicating technical professionals think *of* as they design such new products as apps, mechanisms, and robots.

Ideating on Solutions

In the mindset of design thinking, new ideas are a rich reward derived through collaboration. *Ideating* describes the process of forming ideas. You imagine or visualize possibilities for solving a problem or realizing an opportunity. People with different experiences, disciplinary knowledge, and perspectives work together to generate possibilities and options. In the shift from thinking about tasks to thinking about resources, you and your collaborators are yourselves a major resource.

But you don't just talk. Especially in spaces being designed with such ideating in mind, but elsewhere, too, project collaborators make things. The specialized spaces, called "makerspaces" or "maker studios" or "design studios," encourage innovation through open display of ideas and objects as they are developed. Team members jot down phrases and sketch ideas on white- or blackboards attached to the walls, or on the walls themselves. These marks embody their thinking so that others in the space can interact with it. Some academic makerspaces are designed and created by the student users themselves, who put together inexpensive whiteboards or tables as maker exercises. They own the space. People walking around may also comment on projects whose development is publicly displayed. That kind of serendipity, gained through chance encounters, is a benefit of this design approach. For designing to work, dedicated spaces are welcome but less essential than an open, collaborative mindset and method for swiftly generating ideas. The process works in a virtual environment as well.

For more about collaboration, see Chapter 4

Prototyping

The third phase, not necessarily in sequence but engaged in iteratively with empathizing and ideating, is creating prototypes. A prototype is a model or scaled-down version of your product created quickly and inexpensively to test an idea before you put a lot of energy or money into it. Equipment such as 3D printers and software apps aids in making things.

For more about content management, see Chapter 4

When your product is communication, your prototype is a draft. You assemble the content for prototype drafts through iterations. Write quickly. Think of segments, self-standing, topic-based units of text and visuals that you can assemble in different ways later for different audiences, contexts, and purposes. The trial-and-error mindset is captured in such mottos as "fail harder," or "fail forward," or "fail early and often." You don't look for an early resolution, a "once and done" approach to a draft. You're comfortable with keeping alternatives. You may have to go back to refine your understanding of what the audience really wants and how best to match that need. Review the segments and the draft collaboratively. Redesign the draft as needed.

For more about final design, see Chapter 11

Testing and Evaluating

Through repeated cycles of evaluating and redesigning your draft, often with the participation of your audience, you confirm the best approach, the best idea. Revision—that is, re-seeing—is not something tacked on at the end.

Empathizing with clients and generating alternative solutions to problems have, for example, aided developers in creating public and legal documents, especially instructions and websites. In doing so, developers have moved from a guidelines approach to contextual testing, away from generic actions to situated ones, encouraging the participation of clients and users as collaborators in building the text. The process helps ensure that instructions work for the people who need them. It also helps answer an ethical commitment that public and legal documents be clear and transparent for citizens who must follow them.

CULTURAL DIMENSIONS OF COMMUNICATION

As this description of design thinking should make clear, human values are at the core of the process. Similarly, such values are at the core of empathetic communication. You behave and communicate in principled ways centered on respecting other people. These principles derive to some degree from your upbringing in your culture as you recognize the inherent values and duties your participation in the various communities you belong to requires.

Broadly defined, culture consists of the behavior, social norms, knowledge, habits, values, and language of a group. The group may be an organization, a nation, a tribe or ethnic group, or a profession. Recognizing differences in culture helps you better understand your audience. In addition, it helps you look beyond those who are just like yourself to imagine how your ideas and your work will affect people who are underresourced or underrepresented.

High-Context and Low-Context Cultures

Several years ago, the anthropologist E.T. Hall developed a system for understanding the communication implications of cultural differences that has helped many communicators build a shared context through empathy with diverse audiences (see box).

HIGH-CONTEXT AND LOW-CONTEXT CULTURES

High-context cultures:

- Homogeneous, with fairly strong distinctions between inside and outside.
- Group oriented. Individuals identify themselves as members of the group.
- Unwelcoming of deviant behavior. The culture has high expectations for individuals to internalize group norms and behave accordingly.
- People oriented. The focus is on maintaining relationships.
- Nonconfrontational. Business requires soft bargaining and indirection to preserve face.
- Humanistic. Control is internal, preprogrammed in each individual as part of the culture.
- Averse to risk.

Low-context cultures:

- Heterogeneous and generally open to outsiders.
- Oriented to the individual. Individuals seek to fulfill their own goals.
- Welcoming of a variety of behaviors; you "do your own thing."
- Action and solution oriented. The focus is on completing the task.
- Able to separate a conflict from the conflicting parties. Conflict in pursuit of a goal is positive. Business requires hard bargaining with a direct, confrontational attitude.
- Procedural. External rules govern behavior.
- Welcoming of risks.

Source: Adapted from Ting-Toomey (1985)

In high-context situations, speakers or writers can leave things unsaid or use highly coded language, knowing that they share with the audience the prior knowledge essential for interpreting the message. You and your friends, for example, probably can use shorthand expressions that require only an emoji, a few words, or just a certain kind of look to communicate. Low-context situations bring together a more heterogeneous group of people who do not share the same common ground or base of knowledge. This lack of common ground has many implications for communicators, one of which is that more information needs to be recorded in explicit, detailed messages.

For more about creating common ground in a document, see Chapter 8

Moreover, levels of trust in society also run from high to low. In high-trust societies, people share a vision that encourages collaboration and allows for informal and unwritten contracts. In low-trust societies, people are less able to agree or solve disputes without resorting to written contracts or litigation.

Audiences in the global community are complex and multifaceted. Attempts to categorize culture traits like this simply provide a quick way to analyze an audience through a general understanding of their cultural traditions before getting to know them as individuals in a specific situation you share.

Organizational Culture

The distinction between high and low context can also help you understand the culture of the organizations you work for or belong to. Learning about an organization's culture is an important part of your preparation as a job candidate. Such knowledge continues to be significant throughout your career. To learn about an organization's culture, for example:

- Review its website for direct statements about its mission and values.
- Note any sponsorships, endorsements, or causes the company supports.
- Read its social media channels and articles about it.
- If you visit the company, be alert to the design of those physical spaces.

- During the visit, pay attention to how you are greeted and how others move in the space.
- Ask about culture in your interview, for example:

  o "How do you measure success for someone in this role?"
  o "What do you like most about working here?"
  o "If you could give someone one piece of advice about working here, what would it be?"

If you like what you've learned about the company's culture, then in your interview talk about why you'd be a great fit within that culture, in that work environment, and how you would thrive there (Acosta, 2021).

An International Culture

Similar technical training and education, interests, organizational roles in multinational companies, and entrepreneurial pursuits as well as consumer tastes are making professionals around the globe think, and sometimes even look, alike. People worldwide often share their preferences in social media, may be fluent in a common language (English most commonly), are comfortable conducting their work through the same enterprise collaboration software and video conferencing platforms, and travel through airports that tend to look alike. Those similarities may suggest that they share your communication context as well. But in the 21st-century global community, with work and knowledge workers distributed across the globe, differences remain. You often need to build that context by first understanding differences in culture and then benefitting from those differences as a springboard to innovative thinking. Through cultivating difference, you'll find ways to solve problems and uncover opportunities in your community.

ETHICAL DIMENSIONS OF COMMUNICATION

As a technical professional, you'll be engaged in the large picture of adapting technology to people and people to technology. You'll do so in rapidly changing and often uncertain situations. To do so ethically, you limit your actions and your self-interest to serve a community of others. The term "community" may seem vague or overused, but it provides a good shorthand for thinking in general about how you connect with and respect others. The strategies you will learn in Chapter 3, which focuses on communicating ethically and professionally, will serve you in a range of situations from those where explicit laws and codes of conduct guide your decisions to those in which you need to deduce or negotiate appropriate norms for behavior in line with fundamental ethical approaches.

In brief, ethical communicators respect other people. They are honest, fair, and trustworthy. They fulfill obligations. They respect the law. They are civil and polite. These attributes are easier to state than to put into practice, however, especially in the murky and complex situations professionals often face. Situations you will read about in this book will help you align your development and delivery of a communication product with principles of ethical communication.

DESIGNING IN ACTION: IMPROVING ACCESSIBILITY AT AN AIRPORT

To get a sense of design-thinking-enhanced communication in action, consider this scenario. Travel agents and others advising international airports are increasingly attentive to the needs of a growing and prosperous travel demographic: older people. The manager of operations at an airport, aware of this situation, circulated a request for proposals (RFP) to regional architecture and design firms soliciting their ideas for making the facility more accessible. The manager reviewed these and selected one firm, which assigned a young designer to lead the project. To help her empathize with the physical challenges of an elderly person navigating the airport, she put on a heavy "age simulation suit," as well as goggles and headphones that impaired sight and hearing. She then walked the route from curb to gate (Weed, 2019). She quickly became exhausted. That gave her a better understanding of the problems her client needed to address to serve its elderly customers. She conducted further research, including on-site observation, investigations in print and online sources, and conversations with experts, both in her firm and elsewhere.

Defining the Problem

Her research uncovered known ways to accommodate the elderly, such as benches for frequent rest stops and the elimination of tripping hazards. But two other findings opened possibilities for more innovative solutions. One was that older people tended to look down while they were walking. So, for example, they missed directions given in signs above their heads, where most signs currently appear. A second was difficulty hearing announcements in noisy airports, something she experienced, too, in her age simulation suit.

Ideating

Those findings focused the definition of the problem of "accessibility" to two particular matters: wayfinding and the audibility of announcements. That focus also led to further questions and a need to examine other resources. In this phase, the project team ideated—that is, generated ideas for making the situation better. The client had set the project in motion with a question: how can we make our facility more welcoming to older people? Through empathy with her client's customers, an empathy fostered in part empirically—that is, directly on the ground in her age simulation suit and through additional observations—the researcher further refined her understanding of the problem.

Refining the definition focused the team's approach to gathering further information, especially in sources beyond those in her own discipline of design. These included studies on changes in behavior that occur as humans age and on broad issues of accessibility to those with compromised mobility and hearing impairments. She talked with her client about complaints they may have received from older customers and the results of informal in-house research on the problems of access. She talked with her colleagues at the firm, including space planners, graphic artists, and engineers. The information she uncovered corroborated her two findings as critical. Once accepting the definition of the problem, the designer and her colleagues developed a range of possible solutions.

To do so, they made their thinking open and tangible by jotting down phrases and drawing sketches quickly on a whiteboard in the open, collaborative environment of their office. They also used a folding screen on wheels as a surface on which to tack up the floorplan of the client's airport, as well as photographs of the airport's signage and examples of better airport plans and graphics they found in the literature. An engineer posted a picture of his grandmother to keep everyone aware of the ultimate user community. Over the course of several days, they talked, designed, erased, reconfigured. What they sketched, wrote, or tacked up established the team's common ground. They saw each other's work as they fulfilled mutual responsibilities to the project, building trust and gathering momentum and motivation. Others not on the project came by and made comments as well.

In general, they thought the problem of looking down might be addressed by putting more information closer to the ground. Avoiding shiny floors that might appear to be wet and thus cause people to worry about falling was another potential solution. Technical adaptations to accommodate losses in hearing they considered included introducing systems that transmit announcements directly to a user's hearing device. In addition, they thought that better positioning of check-in desks and greater attention to graphics for wayfinding, both inside and outside, might make the airport more welcoming to every visitor.

Prototyping and Testing

The team then took two steps. One was to prototype the physical redesign with a 3D printer that created a mock-up of the new layout. As a partner in a major firm argues:

> Testing things at a larger scale changes the dialogue with fabricators and contractors. It's no longer looking at a design idea; it's looking at a fabricated object. That changes the dynamic . . . You're having conversations about something real.
>
> (Tomasula, 2019, n.p.)

Such conversation enhances collaborative visualization, gives an early reading on design problems and potential solutions, and allows everyone involved to have a voice in how a problem's solved. Some people comment that, if a picture is worth a thousand words, a prototype is worth a thousand meetings.

The second step was to prototype the communications necessary to complete the project. Documents were needed to gain agreement on the proposed improvements within the firm. The team then needed to submit their proposal to the airport manager, who in turn passed it along to others reporting to him for their approval and to his supervisor at the management company responsible for several airports nationwide. Before the proposed changes could be implemented, the design documents also had to be approved by various government agencies that oversee compliance with regulations, including those concerning accessibility, for government-funded facilities (the airport is managed by a commercial entity but owned by the city). The design documents, too, had to address the needs of contractors and external graphic designers, subcontractors who would implement the proposed changes.

For more about documents addressing multiple audiences, see Chapter 2

All these audiences would play a key role in how the team prototyped not only its design recommendations, but also the delivery of those recommendations in communication products. Thinking down the line to final communications made its process more efficient. It helped the team determine both what to say—that is, the content it needed—and how to say it. The team used a form of workplace collaboration software especially developed for architects that included a messaging function and an application for sharing large design files. The software eased communication within the firm as well as with external audiences. But the usefulness of the software depended on the team's original and ongoing empathy with the client who stood to benefit from the design team's work.

THINKING LIKE A DESIGNER

Thinking like a designer can help you think like a communicator and create communication products that work for diverse audiences in a multimodal environment. This chapter has briefly introduced the process. In the chapters that follow, you'll learn detailed strategies for putting the process in action.

Design thinking can be categorized in five phases, but is essentially an iterative process of connected activities. In Phase 1, you observe and empathize with others as they interact with their environments. Think of ways to make that interaction better. From your thinking emerges a project (Phase 2) in which you collaborate in a team to define the specific problem that needs to be solved or the opportunity for innovation to be developed. Next, generate ideas and gather resources for the project, that is, ideate on what's needed (Phase 3), while you also prototype your solution (Phase 4). One prototype embeds the technical solution. A second is the draft of the communication products needed to accompany your work and deliver it to a client, sponsor, or customer. Revise and re-see that draft (Phase 5) iteratively during the process and in a final design review at the end. Throughout your work, align your draft and the technical product with ethical principles that recognize and respect the cultural values—that is, the human values—of everyone you collaborate with and that your product is meant to serve.

CHECKLIST: COMMUNICATING BY DESIGN

1. Think like a designer as you create communication products
     Center on human values
     Collaborate as you welcome different perspectives
     Learn from your experiences, "fail harder"
     Write quickly, review, redo

2. Empathize with your audience
     Match their level of expertise and interest
     Adjust your communication to other communications on the issue

Determine your shared purpose
To explain so that they understand
To persuade so they can decide or act as you intend

3. Define both the technical and the communications problems initiating your project
Take a conventional approach to a routine situation
Think in terms of tasks
Answer a specific audience request
Incorporate the conventions or specifications appropriate to the situation
Imitate the structure and approach of the appropriate genre

Take a more customized approach to a creative opportunity
Incorporate new technology for assembling a message
Take advantage of the multimodal environment for delivering your message

4. Ideate to generate new ideas
Think less in in terms of tasks than of resources
Consider your client, sponsor, or customer an agent with you in achieving a good result
Review the literature for information as well as models to imitate

5. Prototype, test, and evaluate
Think of a draft as a prototype
Write quickly and share openly

6. Accommodate cultural differences in the global community
High- and low-context cultures
Dimensions of cultural differences
Organizational culture
An international culture

7. Communicate ethically, with respect for your community

EXERCISES

1. To find out about communication products and practices in your discipline, interview a professional practitioner or a faculty member You might ask about the importance of communication skills to career advancement. Then develop questions based on the picture of design-thinking-enhanced communication presented in this chapter. For example:

- How much of your work centers in collaborative projects?
- How do you find out about the people you communicate with and for?
- Do you follow a specific process for talking about meeting their needs?

- Do you have a set procedure for managing a project?
- What workplace software do you use to document and advance your collaboration?
- What conventions or genres of presentation do you use?
- When you communicate across cultures, what practices do you follow?
- Does your organization or discipline offer explicit advice on communicating ethically?

At your instructor's direction, be prepared to discuss your findings in a brief oral presentation to the class.

2. Design thinking has deep roots in several disciplines, but more recently has been codified as a problem-solving process by the d.school at Stanford University and its corporate offshoot, IDEO. Take a close look at their websites to extend your knowledge of this mindset and method. Enter "design thinking" as a key word into your favorite search engine for further information about how this process might work for you not only as you communicate, but also as you solve problems more broadly during your student years and in your career. For example, look at this website to see how it differs from machine learning: https://bit.ly/3nwoEJa

3. As you begin your course in technical communication, write down the goals you plan to achieve during the term. Consider, for example, the strengths and weaknesses of your current communication skills in various modes and media: talking, listening, writing, using various digital techniques for creating and presenting images, videos, audio files, and the like. How wide is your range? What do you need to do to maintain and enhance strengths and overcome weaknesses? Review your performance in the class regularly against this statement of your goals.

FOR COLLABORATION

Organizations depend on professionals who can communicate across cultures. E.T. Hall's description of high- and low-context cultures has been amplified by many other attempts to classify cultural differences in perception and behavior. Enter the names of such researchers into a search engine to discover some of these approaches: for example, Joseph Henrich, who defines a category of WEIRD people (Western Educated, Industrialized, Rich, Democratic) as opposed to those with tribal perspectives. Another often cited source is Geert Hofstede, who defines six (or so) cultural dimensions. In their article "Cultural Intelligence" in the *Harvard Business Review*, Earley and Mosakowski present a simple test to diagnose one's cultural intelligence (CQ), including three sets of statements for use in determining your CQ. In a group of three, take the test and then compare your results. At your instructor's direction, present a brief oral report on your collaborative findings to the class.

REFERENCES

Acosta, D. (2021). Pros weigh in on what to ask in a job interview. *Wall Street Journal*, March 15, A12.

Brown, T. (2020) Design thinking—IDEO. Retrieved from https://designthinking.ideo.com/

Earley, P.C. & Mosakowski, E. (2004). Cultural intelligence. *Harvard Business Review*. Retrieved from https://hbr.org/2004/10/cultural-intelligence

Hillier, B. (1996). *Space is the machine: A configural theory of architecture*. Cambridge, UK: Cambridge University Press.

Ting-Toomey, S. (1985). Toward a theory of conflict and culture. In S. Gudykunst & S. Ting-Toomey (Eds), *Communication, culture, and organizational process* (pp. 71–86). Beverly Hills, CA: Sage.

Tomasula, K. (2019). Is a fab lab in your firm's future? *AIA Architect*, December 3. Retrieved from https://bit.ly/3n3tsoQ

Weed, J. (2019). Accommodations for older travelers. *The New York Times*. April 23, B7.

2 Communicating with Diverse Audiences in a Multimodal Environment

Design thinking begins in empathy with customers, clients, or others, people whose needs, current or potential, initiate your project. In a similar approach, you begin to design and implement communication strategies by empathizing with your audience. You have a purpose in communicating, and they will create meaning from what you say based on their own purpose, prior knowledge, and expectations. Knowing how they'll use the content you communicate, recognizing the role they will play along with you in your shared communication situation, will help you imaginatively ask, "will they understand this term," or "does this need to be explained," or "could they take this the wrong way?" Bringing in users as actual collaborators rather than imaginary ones throughout the project is even better.

In this chapter, you'll first learn strategies for understanding your audience by taking their perspective on your shared communication situation and generating information about them. That understanding will help you meet their expectations as you compose communication products for them, the subject of the next section. In your career, you'll also address the needs of multiple audiences in a sequence both inside and outside your organization, as well as audiences internationally. The final sections of the chapter address those situations.

Parts 3 and 4 of *Designing Technical and Professional Communication* provide detailed advice on composing communication products that respond to a wide range of occasions. Some are relatively common. Some are more novel circumstances that call for innovative and entrepreneurial thinking. This chapter sets the stage with its focus on audience.

TAKING THE AUDIENCE'S PERSPECTIVE

Your communication has to work for your audience as well as for you. As you define the communications problem or opportunity that accompanies your project, develop your understanding as well about the way your audience is looking at the situation you share (see box). Here are some examples:

1. A technician at a company that manufactures kitchen ranges is asked by his supervisor to test the surface heat of their cooktops and compare them with models from their competitors.
2. An interdisciplinary team of mechanical engineering, biomedical sciences, and business students prepares a "concept business plan" to develop a mechanical

DOI: 10.4324/9781003093763-3

system to accommodate adaptive users of a rowing shell. The plan responds to a university-wide competition offering start-up funding for innovative ideas.

3. A group of students, selected from volunteers representing different demographics and interests, forms a task force to encourage compliance with standards of safe behavior on campus in response to the COVID-19 pandemic. They develop a poster to be reproduced online and to be posted at significant locations on campus. The poster is part of a brand package for the initiative, with the logo, "Protect the Flock," reflecting the university's mascot, a Blue Hen.

The technician and the two teams develop and structure information to meet their own purposes and the needs of different audiences:

- A supervisor at work
- An interdisciplinary review committee of faculty and local community members
- Students and others in the broad university community.

Each communication product has a different purpose:

- To explain in answer to a request
- To explain in support of a decision on funding
- To persuade others to act in a desired way.

In composing their different products, the teams look into their audience's purpose, prior knowledge, and expectations.

GENERATING INFORMATION ABOUT YOUR AUDIENCE

Purpose:

- What is the communication's purpose?
- Why should the audience pay attention to this communication?

Prior knowledge:

- What do they already know about the topic?
- What is their general level of expertise in the field?

Expectations:

- What do they expect from you? Are you credible?
- Do you need to establish credibility by creating an effective persona?
- Has the audience given explicit instructions for format and content you need to follow?
- What genres and conventions apply in this situation?
- Does your audience share an understanding of these genres and conventions?

- Do you need to adjust your arrangement of material or teach them how to use your communication?
- What media will your audience find most comfortable for the delivery of your message?

Addressing multiple readers:

- Do you need to address multiple readers with one form of communication?
- If so, what is the relationship between or among the readers?
- Who is the *primary* reader?
- Are there *immediate* readers in your organization who must approve the communication?
- Are there also *secondary* readers who will be impacted by the message?
- Have you designed and labeled sections your communication for easy access by different audiences?

Purpose

You communicate to achieve some purpose through other people, who in turn have a purpose in using what you are communicating. One way to create an understanding of your shared purpose is to consider which of two general categories of purpose—explaining and persuading—fits the situation. In addition, to the degree that you lack close connections with your audience, generate information about them and their own potential purpose you're addressing.

Explaining and Persuading

To simplify, you communicate to *explain* when you want the audience to understand something or to enable them to do something:

- To answer a question
- To see the logic of a procedure or plan, such as a business concept plan, that's familiar to you but unfamiliar to them
- To help them use a tool you know how to use.

By requesting an explanation, the audience signals its willingness to pay attention to what you have to say.

More difficult are situations in which an audience doesn't recognize a need that you see or sees things differently from you. You must first attract their attention and then *persuade* them to engage with your message. You need to explicitly create an overlap between your purpose and theirs. You communicate to persuade when you want your audience to, for example:

- Buy your product or service
- Hire you
- Act in a certain way

- Change their mind
- Come to a decision you think is right
- Agree that something you think is significant is indeed significant
- Approve the course of action you'd like to undertake
- Invest in your business plan.

Your audience will use your communication as a tool to get their work done or as advice for improving their lives or well-being. To return to our earlier examples:

- The supervisor reads the report on cooktop temperatures to understand current conditions before responding to a complaint from customers.
- The review panel reads concept plans from several teams as supporting documentation before panelists listen to each team's pitch about why it should win the competition.
- The student committee, representing different campus demographics, reviews suggestions from its members and from a campus-wide survey to create a persuasive campaign.

People often act in their own self-interest, may resist change, and may resent and challenge opposition. You have to frame any change as less threatening than they may think and call on higher values or interests as necessary. That's why, for example, campus administrators call on students themselves to volunteer on a committee to advocate with their fellow students. They share similar interests, tastes, and social media preferences. The administrators feel the students are in the best position to persuade their fellow students.

Generating Information about the Audience

Lacking ready-made bonds with your audience, you have a range of ways to find out about the people you are writing to or for. In a professional context, with people you don't know, search for information about them on a search engine such as Bing or Google. Take time to do your audience homework. The more significant the message you need to send, especially when you might be asking for a favor from them, the more important it is for you to learn publicly available information about them. Reduce as much uncertainty as you can in the communications situation. If you know you and your audience have mutual connections, seek information from those connections. Avoid making easy and potentially embarrassing mistakes that distract the audience from your central discussion.

When you write to change behavior, as when you are addressing a public audience as part of a social or environmental project or are giving health and medical advice, be particularly aware of culture differences and ethical concerns in attracting their attention and crafting your message. You'll read much more about explaining and persuading in Chapter 8.

Prior Knowledge

At this point in your studies, you are probably most skilled at communicating with audiences who share your disciplinary interests and knowledge and make clear what they are looking for. Consider the specific advice you have received for writing a variety of reports tracking your activities in a class or a lab. Instructors provide models and specifications for these documents, and you adhere to them. You assume that your audiences (professors, teaching assistants) are skilled at reading about the subject. It's their job to read you.

From Experts to Novices

But even as a student, and later as a professional, you'll need to address a range of audiences, from experts (who are well informed before they read) to novices (who know little). Writing for expert colleagues, who share a high-context environment with you, encourages shorthand, largely a shared technical language and a high level of abstract thinking. To the degree that your audience is less familiar with your topic, you need to introduce elements to fill the gap. You'll be less theoretical and more concrete, with examples and visuals created to explain. You may place the focus on real-life applications of a new theory. You'll define critical terms.

Increasingly, researchers are using stories told both in text and in visuals to capture attention and engage interest in scientific and technical matters, especially in communications appealing to the public. Stories have a dramatic pull that can create a desire to learn more about a subject and to answer the non-scientist's question, "What's in it for me?" Framing a narrative about an innovative product or service is a particularly important skill for entrepreneurs.

From Familiar to New

For any communication product, sort through the information you need to present to determine what will be new to the reader, and thus what you need to explain and elaborate on, and what the reader already knows. Begin with a review of the known as a framework for the new. For those new to your subject, you may need to spend more time on background and preliminary steps (such as how to locate an on/off switch) that would tax the patience of an expert.

For more about
familiar-to-new order,
see Chapter 10

Prior knowledge can foster understanding. But it may also impede comprehension if that knowledge conflicts with new or perhaps unwelcome information being presented in your document. In addition, ingrained habits of thought may make learning difficult. As chemist and Nobel Laureate Roald Hoffmann notes:

> When I try to explain chemistry to outsiders, I have three main audiences: the person in the street, fellow academics in the humanities and physicists. All three audiences are equally ignorant of chemistry, but the most difficult audiences are the physicists, because they think they understand, but they don't.
>
> (Browne, 1993, p. C1)

UNDERSTANDING THE AUDIENCE'S EXPECTATIONS

In Hoffmann's terms, the physicists let their prior knowledge stand in the way of new information. He is playing a bit with the idea that physicists may be as "ignorant of chemistry" as the person in the street. But he's making an important point that, although you don't want to overestimate your audience's knowledge of the subject, you should also be aware that firmly held knowledge that is contrary to yours can make communication even harder. Your audience also interprets the content you are delivering to them to conform with what they expect from:

- The context, the shared situation in which you are communicating
- You, as an individual or in your role in an organization
- The genre or conventional arrangement of the communication
- The media and mode in which you've chosen to address them.

Context

Workplace communication emerges in a context of previous conversations as well as texts and images, online and in print, that you and your audience share. Some may be private between you and the audience. Others are available in the wide range of sources you and your audience routinely engage with. Compose with all these sources in mind to control the range of possible interpretations of what you say. You can leave things unsaid or use highly coded language, for example, if you know the audience has the prior knowledge of your particular situation essential for interpreting the message. If you're addressing a group of people who do not share an understanding of the situation, you may need to say more, perhaps redundantly, in text as well as visuals and other media.

Your Persona or Voice

Communicating with diverse audiences begins with understanding yourself. First, assess your own interests, perspectives, and role as you generate your understanding of your audience in relation to you. What is your standing with them? Establish yourself as someone whose communications are worthy of their attention. Is your audience well aware of your expertise and authority in the situation, or do you need to establish your credentials? Audiences are more likely to pay attention to a message from someone they know and trust than from the many others who communicate with them.

A major mark of seriousness and professionalism is the *persona* or voice you create through your communication that confirms your authority and dependability. It also engages the audience. More than just content, you convey something about yourself when you communicate. How much personality you reveal and the voice you choose depend on the situation. You'll probably take a relatively impersonal approach in a formal report. But an email, text, or social media posting that is either soberly serious or silly may strike readers as off-putting. It may fail to establish the authority of your words or images. The persona you evoke may be yours personally. Or you may represent the image or character of your organization.

For more about persona, see Chapters 8 and 11

Genres and Conventions

Second, many situations for communicating scientific and technical information repeat themselves. In response, genres and conventional patterns for arranging content have emerged. Take advantage of them. Your familiarity with them helps you select, organize, and deliver content. Your audience's familiarity with these genres and conventions also helps them access what you say. Use your knowledge

For more about
models in the
literature, see
Chapter 7

of the literature in your field to ask the questions readers would ask, to select acceptable forms of evidence for answering the questions, and to sequence that information in the right way.

Although genres and conventions common in North American settings are often applicable more widely in the global economy, they are not universal. Keep that in mind as you assess your communication situation. Differences in both corporate and national cultures often show up as differences in genre—that is, in expectations about how content will be presented.

An American company successfully adapted its instructions for Japanese customers from the genre of the manual familiar to US readers into the genre of the *manga* of comic book, familiar to Japanese readers.

Aware that the Japanese value the artful packaging of food, presents, and flowers, another American company gave its manuals a glossier look, including bright colors and photographs, than those for the American market.

You meet your audience's needs by understanding how they expect content to be delivered to them. That may be relatively easy, as in presenting a manual as a *manga* or adhering to a framework explicitly requested by the audience. But sometimes you need to negotiate with the audience not only about what needs to be said, but also about their preferred format. There's an old saying in marketing that remains pertinent: you can buy in your language, but you must sell in the language of the customer. Consider the audience your customer and adhere to their expectations. Or teach them to expect something new, either in conversations before you write or in an introductory section to your communication product.

Method of Reading

Third in this brief list of strategies for meeting audience expectations, think about how, and maybe where, your reader will experience your communication. Simplify whatever elements you can to reserve the audience's energy for conquering information that is inherently difficult. Audiences are often resistant and inattentive, tossing what seems unsuitable, selecting material worth a second glance, and then looking at sections only in bits and pieces. They scan and skim to find enough information to solve their problem or perform their task. To get them to settle into your message, make the information they most need most accessible to them.

For more about
arranging content,
see Chapter 10

In addition, in print and online, accommodate the direction in which your audience reads. Arabic and Hebrew, for example, are read from right to left (the reverse of English); Japanese is read from right to left and top to bottom. This affects how people read visuals as well. Japanese charts, for example, sometimes include the vertical axis on the right side. Japanese readers of a diagram that included a "man/auto" (for manual/automatic) switch called it an "auto/man" switch as they read from right to left. When you arrange a series of pictures that show a process, consider using a vertical arrangement to avoid confusion of what is the beginning and what is the end. Most people read vertically arranged material from top to bottom unless directed otherwise.

Media and Modes

Select the most appropriate media and mode or modes of delivery from the ever-expanding range of easily available options. Your audience's preferences or abilities may limit the choice, a limitation you need to respect. But you have multiple choices for engaging audiences in new ways to gain their attention and foster understanding. Print remains a reliable medium, welcome by most audiences and often an important default or back-up way to deliver content that needs to last over a long time. Digital media, online or otherwise, are fast, efficient, and cheap, once you (and your audience) have the right tools. It's relatively easy in a digital environment to foster interaction directly *with* your audience, asking questions and chatting, as you compose. If you need to gain empathy for social action, you might embed links to podcasts about the topic. You might record interviews with those impacted by an event or natural occurrence. You post these stories online to persuade others in a target demographic to act in a way that complies with your scientific findings.

In addition to various choices in media, consider, too, the various *modes* of delivery, often talked about in five categories aligned with the five senses:

- Text (in print and digital, that is read or spoken—that is, read aloud) .
- Visual (images, moving or still)
- Aural (sound, including music)
- Gestural (movement, body language)
- Spatial (physical arrangement, proximity).

If your audience accesses content best by seeing, for example, use visual cues such as images, graphs, or charts. Incorporate videos or posters. Your audience learns best by hearing? Use auditory cues, embed links to podcasts, audio files, or songs.

For more about research that incorporates the senses, see Chapter 7

NEWSLETTERS FOR MARKETING

An architect discussing the best channel for enhancing marketing urges her colleagues not to overlook simple email and newsletters. First, because newsletters sent to an email address offer a company full control over the content. Social media platforms, for example, frequently change their algorithms and offerings. A company needs to update postings and match whatever new ways competitors are delivering their message.

Second, although most current and potential clients have an email address, not all spend time on social media. As of 2019, more than 3.9 billion people used email, according to Statista.com, a figure that is expected to grow to 4.5 billion in 2024. By contrast, Facebook, the leader on social media, had 2.4 billion users, followed by YouTube with 2 billion and Instagram with 1 billion. And, where ads on social media may send an audience to the company's website, it doesn't collect information about who the audience is, information needed to convert them into clients.

Third, email messages can be targeted to an intended audience. Social media have a broader reach, especially for B2B (business-to-business) services. But a marketing report notes that 86 percent of business professionals prefer email to social media for delivering business messages. You can personalize your email to have the recipient's first name appear in the greeting. In addition, unlike with social media, you can segment your audience by their specific interests.

Fourth, emails are cost-effective and relatively easy to produce and update. Asking (perhaps on the company's website) for an email address of visitors and sending a newsletter may "begin a conversation." (Adapted from Lee, 2020)

ADDRESSING MULTIPLE READERS

As an individual or on a team in an organizational setting, you may address a sequence of readers who have different interests reflecting their roles and responsibilities within the organization. External audiences may also need to be addressed, including agencies who regulate the organization's activities as well as stakeholders such as investors or shareholders, customers or clients, citizens, and public officials in the community. Figure 2.1 can help you identify and understand these different audiences.

One document might serve all audiences, or you may need to craft different documents. For example, the designer heading the airport accessibility team wrote a recommendation report to the operations manager, her firm's client and the primary audience. But, along the way, several others at her firm had to sign off on it, including the partner in charge of the project, as well as colleagues in human resources, IT, and graphic design. She had to negotiate with these immediate audiences as they, in turn, became co-authors. Once the operations manager approved

2.1
A Sequence of
Audiences

Source: Adapted from
Mathes and Stevenson
(1991)

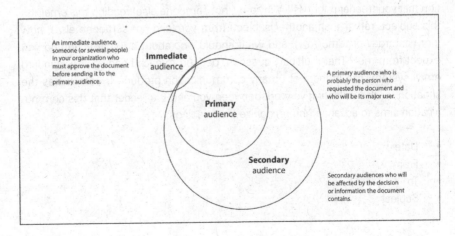

An immediate audience, someone (or several people) in your organization who must approve the document before sending it to the primary audience.

Immediate audience

A primary audience who is probably the person who requested the document and who will be its major user.

Primary audience

Secondary audience

Secondary audiences who will be affected by the decision or information the document contains.

the report, several other audiences would come into the picture, including contractors and regulatory agencies concerned with regional planning. The patient record in a hospital is the major way health care providers communicate with one another. But some hospital administrators insist that the record be written for another audience—that is, lawyers, insurance companies, and review boards who monitor the hospital's accountability.

Depending on their roles and responsibilities, some audiences focus on the technical discussion, others on financial considerations, still others on management issues or policy implications. The path through multiple audiences can become complicated. You may need to create a different document for each audience or design a single document to be easily accessed in clearly designated segments. You may need to negotiate with immediate readers if they contradict each other in reviewing your document, a not uncommon process.

COLLABORATING WITH YOUR AUDIENCE

Think of your audience as a collaborator in accomplishing your joint purpose. The headings in an email message are a good reminder of the major elements you need to address. They derive from the traditional memo, the workhorse document for years when people communicate within an organization. Although much else has changed, those headings still work.

For more about memos, see Chapter 15

Emails today travel worldwide from and to people within organizations. But they also connect people in a network across traditional enterprises. In effect, sending an email is less about fulfilling the conventions of a particular genre and more about the larger picture of sending a message digitally. Such messages tend to be brief, as you respond to requests and maintain relationships. In addition to its many advantages as you compose a message, the digital delivery that defines an email also enhances your ability to be timely with your information. Among other topics, emails report on research or field investigations, describe policies and procedures, and circulate short summaries of meetings.

Structure

When you write an email that is more than a mere comment, think about this question that your reader might ask: Why am *I* (not someone else) reading *this email* (on this subject, rather than another subject) from *you* (and not someone else), *now* (not yesterday or tomorrow), and what should *I do* about it (what action do you expect from me)? That multipart question, pegged to each heading in a standard email system, can in fact set up any communication product. It establishes the situation for the collaborative work between you and the reader that this communication aims to advance. You recognize the headings:

- Date
- From
- To
- Subject

Date

Your system for sending emails probably provides this information automatically, down to the minute. Timing is essential. You write when the information you are sending meets your reader's current need. Emails are fleeting documents, responding to a particular issue, at a particular time. But they are also permanent in creating a record for future reference. If you write when you are angry, or say something indiscreet or without proper consideration, that email may establish a record that later embarrasses you. Similarly, if you write when your reader isn't ready for the information, or is perhaps preoccupied with some other problem, your message may be ignored.

To

You may have one primary reader who requested your email and whose needs structure your response. In your response, you may need to accommodate secondary readers who will want to be aware of what you're saying. You *copy* them, for reference. In addition, you have probably sent, and certainly received, emails that address a group of people: all employees, all collaborators on a project, all people who attended a meeting. When you write to a set of individuals, be sparing in selecting them. You may be tempted to include many, but the more people named on a distribution list, the less likely it is that any one of them will read the memo, because each may think the other is taking care of the issue.

From

As with the date, your system will identify you as the sender of the document. That may be enough identification if your reader knows you. Otherwise, you may need to establish your credentials within the message itself.

Subject

Make the specific purpose of your email clear in the subject line. Here's a simple rule: one email on one topic or action. That's it. When you change the topic, write a new email. Announcing that topic or action in the subject line helps the reader decide how (perhaps "if") and when to read the memo. In the thread of project documentation represented by emails, each category of information—for example, costs, components, personnel—may find a home in a different file. Moreover, each action may require a different approval from immediate audiences. Covering several topics in one email may make responding, acting, filing, and approving difficult.

Greeting

When you write to someone you know well, you may omit a greeting and move directly to the message. When you write to someone you know casually, use their first name, being careful to spell it as they do, especially with shortened names. If your email addresses someone you don't know and is more formal, then you may use the typical letter greeting, for example, *Dear Ms. Jones*, followed by a colon (:). Greetings vary, depending on your relationship and style.

To: CSA@stephens.com
Subject: Cooktop temperatures

Experimental

Chris,

Two experiments were performed to measure the freestanding range cooktop temperature. One was to measure the effect of various burner bowl designs on cooktop temperature. The burner bowls consisted of current bowl and trim, one-piece bowl, and one-piece bowl elevated 1.5 mm. The other was to compare Stephen's cooktop temperature with our competitors'. The competitor's consisted of Freitag, Inco, and Johnson. All of our competitors have trim and bowl and coil heating elements similar to ours. Temperature measurements were taken at various locations. For the burner bowl variations, the tests were repeated to get an average value. The experiment details, including procedures, are in the attached data sheets.

Results
The effect of various burner bowl designs on cooktop temperatures is shown in the data sheets and summarized below:

Cooktop temperature

	Average (C)	Std Deviation
Current	96	21.5
One piece	98	23.0
One piece elevated	112	32.3

The current (trim and bowl) cooktop has the lowest average temperature, 96°C. followed by the one-piece bowl, 98°C, and the one piece elevated, 112°. Elevating the one-piece bowl 1.5 mm not only caused the average temperature to rise by 16°C, it also caused a wider spread in temperatures as shown by the larger standard deviation. With the elevated one pie, the cooktop has a high of 184°C and a low of 73°C. The current bowl and trim have the least spread in temperature. The one-piece bowl design increased the cooktop temperature slightly, about 2 degrees.

If we divide the cooktop into hot central and cooler outer regions, the current trim and bowl and one piece bowl spread is still about 3 degrees in the hot zone, i.e., 116°C versus 119°C. However, the elevated one piece bowl increased the cooktop hot zone temperature significantly to 142°C.

Here is a summary of the comparison test results (see attached data sheets)

	Average	Std Deviation
Freitag	98	15.3
Stephens	96	21.6
Inco	99	21.8
Johnson	80	14.8

Johnson has the coolest cooktop, about 16° cooler than Stephen's. It also has the most even temperature cooktop. Inco has the second coolest cooktop, but it has the highest temperature spread. Stephens and Freitag have the hottest cooktops. We have a bigger spread in temperature compared to Freitag, that is, hotter in the center but cooler in the outer ridge.

About 3 percent of the customer instruction service calls are due to the hot cooktop. There are many calls and letters from consumers complaining about our hot cooktop as well. Our cooktop is so hot that one customer called in to the hot line to complain that her fingers were blistered when she placed her hand on the stove top. It is not surprising since water boils at 100°C. To reduce customer complaints as well as addressing a safety issue, I strongly recommend that we redesign our range to reduce the stove top temperatures.

Margin notes:

Vague subject line. What is it about? Descriptive heading is more common mid-document

The email begins without any context or motivation for the reader to read

Details lack connection and emphasis

Approach is that of the author's lab notebook not a reader-oriented message

Paragraph restates the details without answering the central question in the consumer complaint

Facts are hard to compare and paragraph doesn't develop toward a point

Important recommendation is buried

**2.3
Revised Email
Report**

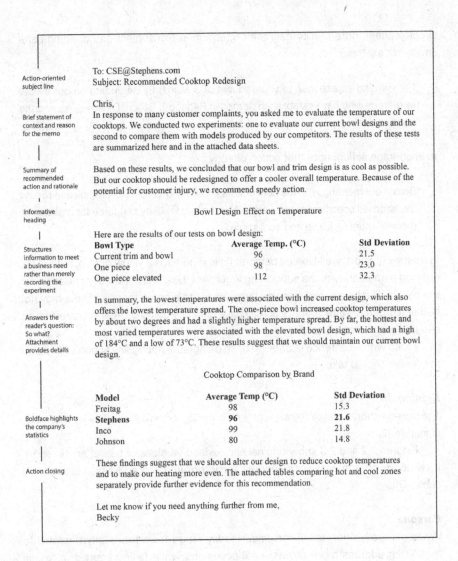

Action-oriented
subject line

Brief statement of
context and reason
for the memo

Summary of
recommended
action and rationale

Informative
heading

Structures
information to meet
a business need
rather than merely
recording the
experiment

Answers the
reader's question:
So what?
Attachment
provides details

Boldface highlights
the company's
statistics

Action closing

To: CSE@Stephens.com
Subject: Recommended Cooktop Redesign

Chris,
In response to many customer complaints, you asked me to evaluate the temperature of our cooktops. We conducted two experiments: one to evaluate our current bowl designs and the second to compare them with models produced by our competitors. The results of these tests are summarized here and in the attached data sheets.

Based on these results, we concluded that our bowl and trim design is as cool as possible. But our cooktop should be redesigned to offer a cooler overall temperature. Because of the potential for customer injury, we recommend speedy action.

Bowl Design Effect on Temperature

Here are the results of our tests on bowl design:

Bowl Type	Average Temp. (°C)	Std Deviation
Current trim and bowl	96	21.5
One piece	98	23.0
One piece elevated	112	32.3

In summary, the lowest temperatures were associated with the current design, which also offers the lowest temperature spread. The one-piece bowl increased cooktop temperatures by about two degrees and had a slightly higher temperature spread. By far, the hottest and most varied temperatures were associated with the elevated bowl design, which had a high of 184°C and a low of 73°C. These results suggest that we should maintain our current bowl design.

Cooktop Comparison by Brand

Model	Average Temp (°C)	Std Deviation
Freitag	98	15.3
Stephens	**96**	**21.6**
Inco	99	21.8
Johnson	80	14.8

These findings suggest that we should alter our design to reduce cooktop temperatures and to make our heating more even. The attached tables comparing hot and cool zones separately provide further evidence for this recommendation.

Let me know if you need anything further from me,
Becky

Paragraphing

For more about the
important role of
paragraphs in design,
see Chapter 10

Following the header, provide the email's main content. You don't need to create a five-paragraph essay in an email, the writing approach you may have learned earlier in your studies. But you do need paragraphs for anything more informative than a mere comment. The indention for the paragraph helps move the reader into your message and provide direction though the various segments of your explanation or argument. Each paragraph may be only one or two sentences long. That's fine.

Use a brief opening paragraph to make clear what your subject line implies. Set up the situation for your email by answering the reader's question about why you are writing to them and what they should do about it. The extent of your explanation depends on your previous shared knowledge with the reader. If you are uncertain about their knowledge, err on the side of being explicit. If you're responding to a direct request, note that to gain their attention in the subject and opening paragraph. For example:

> I am writing in response to your request of 3 March for an update on our survey of retirees who have established homes in Rehoboth Beach, Delaware, during the last five years.

To encourage action, note that action directly:

> Before our meeting with the consultants on Friday (2 pm, via Zoom), please review the attached specifications developed by the accessibility taskforce for making our website more inclusive and accessible.

In a series of short, well-labeled paragraphs and perhaps a visual or two, follow this opening paragraph with the supporting argument. Use attachments, as needed, for the details. But be aware when you send emails to a group that recipients may not bother to open the attachments or may have difficulty doing so. Any key information should be in the email itself.

Close with a sentence (a one-line or so paragraph) noting any action or decision the reader should take.

Signing off

The sign-off should match the style of your greeting, often followed by an automatic signature block.

Figures 2.2 and 2.3 show original and revised versions of a brief email report. It responds to a supervisor's request for information to address a customer's complaint.

Persona

While generally brief, email messages today range from the highly personal to those that address more professional occasions while falling short of a formal report, like the examples in Figures 2.2 and 2.3. An effective assessment of your relationship with the reader or readers in the situation at hand is key to knowing how to pitch not just your message, but also your voice, the persona you present in expressing the message. That voice is prominent especially as you open and close your message.

Taking the Audience's Perspective

For the situation at hand, whether an email message or any other communication product, you determine your audience's purpose in learning what you know, their prior knowledge of the subject, and their expectations from you and for the communication. As you do so, you also help ensure that the communication will work for you as well. You are partners, collaborators, in exchanging information productively to solve a shared problem or pursue a shared opportunity.

This chapter has focused on empathizing with your audience—that is, understanding them by generating information explicitly about them and taking their perspective. To simplify, consider the purpose for communicating you share in two broad categories: explaining and persuading. Assess their prior knowledge of the field and of the topic of concern in this particular situation. Assess their knowledge, too, of common genres and conventions for arranging information you share. Adjust your thinking if you are communicating with others in a different regional or national culture. Recognize situations where your communication must meet the needs of multiple audiences within and outside your organization. Accommodate the special situations that arise in communicating across cultures regionally and internationally. Collaborate with your audience to ensure a good outcome for everyone.

CHECKLIST: COMMUNICATING WITH DIVERSE AUDIENCES IN A MULTIMODAL ENVIRONMENT

1. Generate information about the audience
   - Take the audience's perspective
   - Determine their purpose and yours in communicating
     - To explain
     - To persuade
   - Assess their prior knowledge about the topic and the situation

2. Meet the audience's expectations
   - Identify what they expect in the context of the situation
   - Assess your own role and their expectations of you to create an appropriate persona
   - Match the conventions and genres the audience expects
   - Consider how (and maybe where) they will use the document
   - Select the appropriate media to deliver your communication

3. Write to be understood across cultures
   - Use the structure and style the reader expects when they're different from yours
   - Collaborate with the reader to develop a new structure and style

4. Accommodate multiple readers
   - Rank your readers to know whose needs are most important
   - Chart the sequence from immediate to primary and secondary
   - Make sure you meet the primary reader's expectations.

5. Collaborate with your audience to ensure a good outcome for everyone

EXERCISES

1. To prepare a report for your technical communication class, focus your purpose by selecting an audience who would need to read about the topic. For example, if solar energy interests you, consider who might need information on the topic as part of their work or personal life:

   • Mechanical engineers and materials engineers
   • Architects
   • Government agencies and legislatures who set public policy
   • Manufacturers of hardware for solar conversion
   • Designers who develop solar systems
   • Economists interested in energy cost–benefit analysis
   • Plumbers
   • Homeowners
   • Home financing institutions
   • Zoning boards
   • Tax accountants

   Discuss in class the kinds of information each audience would require.

2. If you hold a job in addition to your work as a student, look at the reporting relationships within the company and to clients and other external audiences. Determine if there is a pattern of immediate, primary, and secondary audiences for particular documents. You may need to ask about audiences in the home office of the company if you work in a branch office or a franchise. Then write a brief report on your analysis. Accompany that report with a diagram like that in Figure 2.1. Discuss briefly the role of each reader and, thus, their purpose in reading documents sent to them.

3. Write a brief (about 500 words) description of an audience you have addressed or will be addressing in a project you are engaged in—for example, a supervisor at work, a professor, a client, or sponsor of research. Use the box "Generating Information about Your Audience" to frame your understanding and your description.

   Describe not only what you understand about the person as an audience, but the sources you used to enhance that understanding. Did you look up the individual online? Did you read articles or other documents written by the person? Did you interview the individual, gaining information directly in a conversation, or indirectly, by observing the individual in their surroundings, either in a physical office or in the background of a video conference?

FOR COLLABORATION

In a team of three, develop a composite picture of *yourselves* as an *audience*. Those who market all kinds of products, especially online, are eager to sell products and services to you. Your college or university has many reasons to address communications to you. So do other organizations that might want to hire you or enlist you as a member. Through the advice in this chapter about understanding your audience, develop a similar picture of yourselves and what you look for in communications addressed to you, including expectations about how such communications should be delivered in a multimodal environment. Write up your composite description in a memo to your instructor.

REFERENCES

Browne, M. (1993). Scientists at work: Roald Hoffmann: Seeking beauty in atoms. *The New York Times*, 6 July, C1.

Lee, E. (2020). Why architects need to build an email list and newsletter. *Architect*. October 6. Retrieved from https://bit.ly/2QFq5sb

Mathes, J.C., & Stevenson, D.W. (1991). *Designing technical reports* (2nd ed.). New York, NY: Macmillan.

3 Communicating Ethically and Professionally

In a simple phrase, the naturalist Aldo Leopold talks about ethics as a "community instinct" (quoted by Ross et al., 2019, p. 8). That means behaving well with others. The term *community* provides a good shorthand for thinking in a broad way about how you connect with and respect others. In doing so, you limit your actions and your self-interest to serve a community composed of different, independent but interdependent parts, human and nonhuman. A community instinct can help you make ethical choices while you communicate in the rapidly changing and often uncertain situations of the 21st century. In this chapter, you will learn strategies for developing your community instincts and for behaving well as a technical and professional communicator. These strategies will serve you in a range of situations for communicating, from those where explicit laws and codes of conduct guide your decisions to those in which you need to deduce or negotiate appropriate norms for behavior in line with fundamental ethical approaches.

LEGAL IMPLICATIONS IN TECHNICAL COMMUNICATION

In the low-context culture of the United States, detailed and comprehensive laws regulate many of the ways that people should respect one another. This includes protecting their privacy as well as their property, both real and intellectual, and ensuring equal treatment and access to services. These laws evolve over time in the various jurisdictions of government as conditions and cultural priorities change. Changes in the law often are needed to respond to new technologies and new social movements. In higher-context settings, you may encounter fewer written guidelines and a greater need to understand the basic values of the culture. From these values, you determine expected patterns of good behavior. As a student, and especially as a professional, you're responsible for understanding the laws applicable where you are pursuing your projects, as well as those pertaining to locations where the product or service you are developing will be used or will take place.

For more about high- and low-context cultures, see Chapter 1

Of particular interest to students or professionals writing about technical information are laws governing the use of someone else's ideas or language, their *intellectual property*. In a report or article, citing sources justifies your stance, indicates what's new in your thinking, and details how your investigation departs from what others have done. Proper citation shows that you know the work of people the audience admires—and that you are thus in good scientific company. Being part of this

DOI: 10.4324/9781003093763-4

For more about
reusing and
documenting sources
in your writing, see
Chapter 7

professional community means that you acknowledge the originator of evidence you provide to support your ideas. To use someone else's texts and images without such acknowledgment is plagiarism.

In the broader picture, legal guidelines establish *liability*—that is, for example, whether advertising is deceptive and whether a product, service, or process has the potential to cause harm. Manufacturers of industrial and consumer products keep a close eye on liability laws when they write manuals and other product documentation. The Product Liability Directive of the European Union (EU), for example, places special emphasis on documentation. It holds manufacturers liable for defects caused when instructions do not meet minimum requirements. Citizens worldwide are also increasingly concerned about maintaining their privacy and control over personal data as they use digital technology. Governments are responding. For example, the EU's General Data Protection Regulation imposes strict measures to remedy circumstances in which technology companies misuse such data. Privacy protection is an often-contentious topic in the US legislative environment as well. In the US and EU, explicit guidelines also define *libel*—that is, false statements that damage someone's reputation or career. Laws also regulate many aspects of workplace practices, including hiring, safe and equitable behavior on the job, and termination.

DEVELOPING AN INSTINCT FOR COMMUNITY

Although legal standards carve out a good deal of territory, they do not cover all situations in which citizens interact with each other, with their environment, and with technology. Laws often engender questions of interpretation and application, locally as well as internationally. In the complicated and murky world of your professional life, and in your personal life, you'll encounter many situations not easily resolved by reference to existing laws.

Social media, for example, provide a broad platform for sharing text and visuals that often poses questions not easily resolved by traditional definitions of plagiarism or by copyright standards. For example, it's easy to share manipulated images. Memes—that is, photos or screenshots with text superimposed—are often doctored. Their origin, including quotations taken out of context, may be hidden behind several reuses or circulation.

More broadly, in any situation, you have to understand the complexities of the relationship between an agent (human or nonhuman), an action, and the consequences of that action. You'll rely on an understanding of the fundamental values and norms of ethical behavior to choose what to do. An instinct for community is a major asset in making those choices.

Community Messaging During a Pandemic

Communities arise out of a felt need to collaborate with others in meeting a common purpose. As a dramatic example of community formation, consider the various responses to the worldwide COVID-19 pandemic that started in 2020. When people are faced with this kind of situation or challenge, something not of their making, they tend to develop communities to provide mutual support. They do so by communicating. Some of that communication consists of messages aimed at persuading others to think of themselves as members of a community and, thus, behave well. A quick look at a few brief messages circulating in 2020 to encourage sheltering-in-place, an action then considered by governments and health professions as essential to containing the virus, can serve to introduce and illustrate three long-standing approaches to making ethical choices. The messages appeared everywhere: on the smart signs over or along highways that usually update information about traffic delays, on public buildings, in parks and recreational areas, in magazines and newspapers, and in social media.

Cultivating Your Good Character

One message went like this: "Stay Home. Stay Safe" or "Stay Home. Stay Healthy." In simple terms, this message illustrates an ethical approach centered on your own values as an individual. You cultivate your good character in a way that allows you to live the best life you can. That includes personal integrity, courage, a sense of justice, and generosity. Your good character, in turn, will help ensure that your actions will be right for others and for the situation. If you stay home, the message suggests, you'll be ok, and you'll help others to be ok.

Seeking the Greatest Good

Another message, "Stay Home. Save Lives," aligns with a second well-known approach to ethical decision making focused explicitly on the consequences of your act. This principle is often stated as seeking the greatest good for the greatest number of people. It means that you can pursue your own good interests so long as you don't deprive others of theirs. You leave others at least as well off as they were before your action. The principle is well established in the physician's code: Do no harm. The message ties together act and consequences: If everyone stays home, everyone will be safe.

Determining if Your Action Is Right

A third version of the message aimed to persuade people to stay home in recognition of the particular right of a special group, health care workers, who served everyone on the front lines of saving lives. The message, "Stay Home. Save Them," was often accompanied by pictures of such workers. It illustrates a third traditional way to frame ethical decisions: determining whether your act is not only good but also *right*. You ask, for example, what duties do I owe? How do I decide between conflicting duties? In this approach, you also act so that you never have to treat others, or even yourself, simply as a means to an end. You don't involve others in a scheme to which they can't consent and, thus, diminish their humanity. You'll read more about these three traditional ways of thinking about ethics throughout the chapter.

A sense of "We're all in this together," an oft-repeated statement in 2020, fostered the growth of various types of communities that emerge as well in less troubled times. These communities often develop mainly online, particularly through social media, to foster common interests in the arts, music, cooking, social action, or other activities. Neighborhood groups form to help neighbors. In 2020, these groups, for example, helped the elderly or others who needed such assistance with

**3.1
"Stay Home"
Messages on
Government
Websites**

Examples from
state government
websites (left to
right): Washington,
Utah, and Oregon

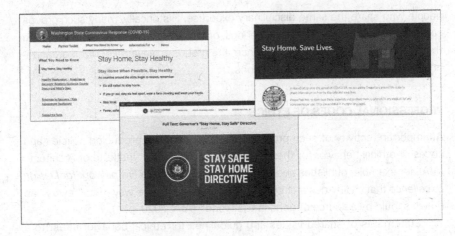

shopping for food or medicine. They posted and exchanged information about best practices in keeping safe, such as obtaining and wearing masks; about available resources and government advisories; and about activities, such as collections for food banks, and events. One such event in 2020, for example, was a neighborhood bell ringing. With appropriate social distancing, people came out onto balconies or into front yards individually, near their home, to ring bells at a particular hour for a few minutes to celebrate the services of health care workers.

Informal and Formal Communities

You will participate in many types of communities, informal and formal, in your career and in your life. They emerge both within and across such borders as geography, culture, discipline, and perspective. The brief discussion of communities that developed in response to the 2020 pandemic suggests one type—an ad hoc group coming together to face the same adverse circumstances.

Short of an adversity, informal communities also grow among people who simply share the same interests. Social media are often the launch pad and accelerator for these communities. The spectrum of such interests is wide, including:

- User groups—for example, software developers, designers creating new forms of typography, or health practitioners who post content to learn from each other
- Support groups empathizing with those undergoing similar diseases or situations
- Hobbyists, celebrity fans, and players of similar games who share their enthusiasms
- Neighbors exchanging information about events and services in the neighborhood.

You can probably add to this list from your own experience.

Some informal communities grow into more formal organizations when they reach a certain size, importance, or longevity. One such formal community you will probably join as a professional is a *community of practice*—that is, a group of people who share the same disciplinary expertise. Historically, many such groups came together through informal gatherings, often in coffee shops or taverns. When they reach a certain size and prominence, the groups achieve recognized organizational status as professional associations.

PROFESSIONAL CODES OF ETHICS

An important activity of such professional associations is producing policies and codes of ethical behavior for their members. The American Institute of Architects (AIA), for example, publishes and circulates on its website a *Framework for Design Excellence* that challenges architects to pursue projects in a way that, " every line drawn should be a source of good in the world" (AIA, 2020, p. 1).

For more about the AIA's *Framework,* see Chapter 11

Statements of shared values and guidelines for ethical behavior attest to a professional association's adherence to high standards in how its members perform their work, the ethical reporting of information in the association's publications, and the profession's social worth. As an example, in its statement of ethics, the American Chemical Society (ACS) notes,

> An essential feature of a profession is the acceptance by its members of a code that outlines desirable behavior and specifies obligations of members to each other and to the public. Such a code derives from a desire to maximize perceived benefits to society and to the profession as a whole and to limit actions that might serve the narrow self-interests of individuals. The advancement of science requires the sharing of knowledge between individuals, even though doing so may sometimes entail foregoing some immediate personal advantage.
>
> (Coghill & Garson, 2006, p. 11)

Among the "core values" of the American Society of Mechanical Engineers (ASME), in addition to promoting engineering education and documenting engineering history, are these:

- Embrace integrity and ethical conduct
- Embrace diversity and respect the dignity and culture of all people
- Nurture and treasure the environment and our natural and man-made resources
- Facilitate the development, dissemination and application of engineering knowledge
- Promote the technical and societal contributions of engineers.

(ASME, 2018).

In its agreement of cooperation with other organizations internationally, the ASME further states that the agreement must "ensure that all ASME members are treated alike, independent of age, race, color, ethnicity, sex (including pregnancy), gender, gender identity, gender expression, national origin, citizenship status, religion, sexual orientation, disability, and veteran or military status."

In an initiative to carry out its mission of "advancing technology for the benefit of humanity," the Institute of Electrical and Electronics Engineers (IEEE) in 2019 published the first edition of a 291-page report, *Ethically Aligned Design: A Vision for Prioritizing Human Well-being with Autonomous and Intelligent Systems.* Aimed at creators, operators, and users of such systems, it was the result of a 3-year broad, international, collaborative endeavor. Its publication online through a Creative Commons license makes it easily and widely available. That wide availability further attests to the institute's goal of raising public awareness of both the social benefits and the potential harms and misuse of such systems. Advancing "values-based design methods," the document outlines eight high-level general principles for creators of autonomous and intelligent systems (A/IS). (The term A/IS is preferred over simply "AI" to include both systems that operate on their own, such as care robots and autonomous cars, and intelligent systems that are capable of learning.)

GENERAL PRINCIPLES FOR CREATING AND OPERATING A/IS

1. **Human Rights**—A/IS shall be created and operated to respect, promote, and protect internationally recognized human rights.
2. **Well-being**—A/IS creators shall adopt increased human well-being as a primary success criterion for development.
3. **Data Agency**—A/IS creators shall empower individuals with the ability to access and securely share their data, to maintain people's capacity to have control over their identity.
4. **Effectiveness**—A/IS creators and operators shall provide evidence of the effectiveness and fitness for purpose of A/IS.
5. **Transparency**—The basis of a particular A/IS decision should always be discoverable.
6. **Accountability**—A/IS shall be created and operated to provide an unambiguous rationale for all decisions made.
7. **Awareness of Misuse**—A/IS creators shall guard against all potential misuses and risks of A/IS in operation.
8. **Competence**—A/IS creators shall specify and operators shall adhere to the knowledge and skills required for safe and effective operation.

Source: IEEE (2019, p. 18)

The report provides extensive discussion of practices internationally needed to put these principles into action.

CORPORATE CODES OF ETHICS

Like professional associations, corporations, museums, academic institutions, and other for-profit and not-for-profit entities document their values and norms of conduct in written codes to be adhered to by employees. All such codes assume compliance with the laws in effect wherever the enterprise operates, at home and abroad, although such compliance is subject to negotiation. You'll pursue the problem-oriented projects at the core of your technical work within the overlapping, and sometimes conflicting, circles of professional and business communities.

In the same effort toward transparency and awareness shown by the IEEE, business organizations often make their statements available to the public. In a 56-page document titled *Achieve More* (2020), posted on its website in several languages and required reading for all employees, Microsoft, for example, details its standards of business conduct. A letter from Satya Nadella, CEO, addressed to "Team," introduces the document. It begins,

> Each one of us shapes our culture through our words and actions. We strive to build a diverse and inclusive culture that embraces learning and fosters trust—a culture where every employee can do their best work.
>
> Making good decisions and ethical choices in our work builds trust in each other and with our customers and partners.

The brochure defines fundamental concepts that underlie ethical behavior. One is values, the "enduring principles that we use to do business with integrity and win trust every day." These are "respect, integrity, and accountability." The second is culture, the "operating framework—who we are and how we behave." Figure 3.2, which reproduces the document's table of contents, serves to summarize the key points covered.

STANDARDS OF BUSINESS CONDUCT

3.2
Table of Contents of *Achieve More*
Source: Used with permission from Microsoft (2020)

DESIGNING ETHICAL USES OF TECHNOLOGY

Because they change social relations and can add a new dimension to the meaning and interpretation of the environment, digital technologies introduce new conditions for ethical choices. The benefits of such technology are real and well known. But as the IEEE guidelines noted earlier recognize, A/IS and other technologies can compromise an individual's privacy and agency along with their health and safety. They can also lead to adverse economic and environmental effects. Many of the now dominant technology companies began with a stated mission to make the world a better place. But not everyone has benefitted.

Privacy and Well-being

A serious concern is the way that technology companies such as Google and Facebook and an increasing number of apps collect as much information as possible about you from which they build a profile of who you are, where you go, what you like. From these data, they can predict your future behavior and potentially manipulate that information for profitable resale. In using the app, you make a bargain: an exchange of information, and personal agency, for the pleasures and ease of using the app. This condition is briefly captured in an oft-repeated line about using social media: "If the product is free, you are the product."

In some circumstances, individuals are no longer *adopting* digital technologies but *adapting to* these technologies by changing their behavior. The technologies lead to changes in how they make choices as a consumer, how they interact with others, and the perspectives they take on society and politics. The app draws their attention and can become addictive. Mixed-reality environments, both virtual and augmented, can be even more addictive. Such technologies for extensive data collection, including apps for facial recognition, help technology companies and governments achieve wide surveillance and, thus, behavioral control over users and citizens.

PREDICTING BEHAVIOR BASED ON SOCIAL MEDIA

An insurance company uses Facebook data to price products differently to different prospective customers. The company found that "people who write in short, concrete sentences and use lists are safer drivers; excessive us of exclamation points suggests recklessness behind the wheel (Wu, 2020, p. 19).

In addition to its potential for compromising personal freedom and self-determination, A/IS has also adversely affected social fairness. In its use during the hiring process in corporations, for decision-making in courtrooms, and even in medical diagnosis, for example, it has been shown to demonstrate built-in biases in terms of gender and race. In consolidating vast quantities of data, A/IS embeds past collective experience in a way that can inform current situations. But its decision-making rationale often is not transparent, is hard to explain, and is not easily adjusted for changing social conditions and norms.

Environmental Sustainability and Social Justice

Recognizing some of these ethical challenges arising from the use of digital and other technologies often leads individuals to organize efforts that advocate for ethical change. Such advocacy communities address, for example, matters of environmental sustainability and social justice—that is, fostering the principle that everyone should have equal access to economic, political, and social rights and opportunities.

As a student and in your career, you may participate in such communities. You may, for example, join a neighborhood group to craft policies for recycling or reducing food wastes, participate in a sustainability committee at your university, spend a work-study period with groups such as Habitat for Humanity or a student chapter of Engineers without Borders, or engage in a service-learning activity with a local not-for-profit. These experiences enhance your communication skills as they also reinforce ethical commitments locally or globally.

An Ecosystem Perspective

One way to look at both environmental sustainability and social justice is seeing people and their environment as living together in a larger ecosystem. The land and the ocean have rights, as do those living on the land, in a system that has its own need to preserve its "integrity, stability, and beauty" (Ross et al., 2019, p. 8). They are independent, but also interdependent, members of your community, not to be dominated but to be valued.

That perspective, you also must know, is not universally accepted. Any efforts to advocate for it, if that's what you want to do, must encourage the participation of groups of people who disagree. Some of that disagreement reflects differences in cultural values, which play a role in both damaging the environment and reducing environmental damage. But disagreements also reflect differences in power among those vying to use these resources. The final section of this chapter offers advice on effective and ethical communication practices in difficult discussions.

An Environmental Scenario

To see how an ecosystem perspective may play out in a real scenario, take the recent example of a high-end, high-rise condo with great views of the water in a coastal US city. It has been built to stand above what research has shown to be currently anticipated levels of potential sea rise and flooding (and, incidentally, above competing high-rise buildings, blocking their views). Its fortification against potential flooding, however, does nothing to mitigate the impact of floods that will occur in the low-lying area. It could also make things worse for neighbors as it directs runoff into the water table. An architect looking at the situation argues that an approach called *resilience* can be one way to respond to climate change. But the approach is also, he argues, "exclusionary and unjust; if we can never stop or contain nature, we will just deflect it—onto those who cannot afford to get out of the way" (Shaw, 2020). In ethical terms discussed earlier in the chapter, it engages others in an activity to which they have not consented and, thus, diminishes their humanity. The architect advocates instead for a focus on "equity-minded climate adaptation, on structural changes that will reimagine new urban futures under climate change . . . [to] protect both the physical environment and the social fabric of neighborhoods" (Shaw, 2020).

Advocating for such broad changes, however, requires collective endeavors. It's easier for individuals and business developers to simply leverage their power to fortify against nature. The very term "resilience," which seems relatively positive, can offer its proponents a way to hide from criticism. Wealthy homeowners in beachfront communities, for example, build walls to prevent coastal erosion. But the walls also disrupt tides and deny public access to beaches. Walls can be built and sand dunes can be restored equitably. But such actions come at a high cost in dollars and in the opportunity to pursue alternative strategies.

Advocating for adaptation rather than resilience requires changes in government policies (such as not paying homeowners to rebuild on flood plains). It necessitates regional and urban planning. A regional plan association, for example, proposed abandoning years of efforts to contain the New Jersey Meadowlands in the US by creating a kind of national park that could change shape to accommodate climate change. Thinking in terms of adaptation rather than resilience can lead to technological innovation—for example, the design of an "adaptable" street with "a permeable paving surface and structural soils to connect the roots of street trees." Rain water would flow into the ground rather than flooding homes and businesses (Shaw, 2020).

Sometimes, traditional thinking can be effectively repurposed to serve a new occasion. In environmental matters, consider the mariner's code for determining who has the right of way on the water: Power yields to sail. Those driving a power boat, having more force and control, must yield to those operating sailboats, powered by nature. Resilience yields to adaptation.

BUILDING TRUST AND CIVILITY

Whatever your perspective on the environment and on the role of government in society, as a technically trained individual, you have a responsibility to consider the ethical implications of your work for the global community as a whole. Applicable laws, the codes of conduct for professionals in your discipline, and the ethical standards of the institution or organization where you are studying or employed provide a baseline. You follow them. In addition, informed by design thinking, you empathize with and advocate for the interests and rights of stakeholders who are impacted by and who impact the environment of concern in your project. That ethical awareness has to be in place at the beginning. It's much harder to make changes further down the road. In the end, the measure of your product's or service's success is not just that it's expedient (easy to do) or feasible (you can do it), but that you *should* do it. It has good consequences for all those impacted.

That said, although the traditional approaches to ethical conduct described earlier in this chapter are widely shared in Europe and North America, they are not universal (nor are they universally observed). The following sections provide strategies for communicating in ways that build trust and respect others. The process begins with understanding others, those who will become your audience. To do so, you may have to build common ground through dialog. Building common

For more about understanding your audience, see Chapter 2

ground is particularly significant when you and your audience differ in culture and perspective and in situations where there is resentment or a lack of trust. Even in situations where you share more common ground, keep aware of the need to communicate so that you maintain and enhance trust and community. That community, too, should be an inclusive one that welcomes diversity and does not humiliate or fail to recognize the worth of other communities as well. At core, it should value the human rights and well-being of everyone

Engaging in Dialog

To build common ground for communication that accommodates differences and a lack of trust, you'll often need to disagree without being disagreeable. Here is advice for doing so in a dialog or conversation. Such strategies don't always work. Sometimes, you just have to walk away from difficult situations. But give them a try first.

For more about negotiating common ground as you communicate, see Chapters 4 and 8

Listen with Your Eyes and Heart

This attitude, a principle of user experience (UX) research, is a good start. Ask the other person what they're experiencing in their own life, what's meaningful to them. Let them, not you, take the lead in a conversation on relatively neutral topics you share, as people.

For more about UX research, see Chapters 6 and 11

Then, ask how they frame the situation of interest. From their perspective, their side of the matter is true. Be willing to see that, too, without thinking of them simply as a position to be countered or a problem to be solved.

In highly charged social or political situations, it's often helpful for participants in a discussion to be limited to talking only about their personal experience in similar situations, without references to sources of their opinions in the news and on social media. Sharing the personal and asking questions about experiences that are fundamental to their beliefs can help to establish trust.

For more about listening, see Chapter 4

Take Time to Respond

Avoid a quick, emotional response. Telling someone they're wrong doesn't provide an incentive to cooperate but may only cause them to dig into their position.

Try restating someone else's ideas in your words to identify similarities and differences. Respect explicitly voiced disagreements to see where ideas can be integrated and adapted. That may satisfy them: They weren't wrong: they only failed to see all sides. Agree on some points while also voicing your own counter concerns.

Consider these questions: What options or information are you missing? How might someone else's framing of an argument be shown to miss options or information? What similarities are there in seemingly different concepts, purposes, arguments? How can these different frames or perspectives be aligned to share strategies and create options for solving a problem?

Assume the Need for Common Ground

Proceed on the assumption that the point of the dialog is not to win an argument but to negotiate a solution fair to everyone. The other person may feel wronged or unjustly treated. Recognize that explicitly: "I think you feel this isn't the right solution for you. Can you tell me why?" "What are you seeing in the situation that maybe I'm not?" Keep the focus on the larger ethical principles you and others share. Make information available and transparent to all parties.

Assess Your Position

In any situation where there are disagreements and conflict and you are advocating for a particular perspective, be aware of your own status and access to power in relation to the other—in particular, your different social and cultural conditions. Consider how you can amplify the agency of those who are materially, socially, politically, and/ or economically under-resourced (Jones, 2016). Consider, too, whether—or how—you can share power with all those concerned.

Reflect

Learn from your experiences in conducting dialogs to apply that learning in future situations. Creating common ground and a shared purpose, giving advice and talking through a problem, these activities help you gain a better perspective. Your concern for others and carrying out obligations based on that concern, even in a small way, enhance your own abilities in changing or uncertain situations.

Composing Text and Visuals Ethically

The advice you've read about strategies for building common ground through dialog works as well when you turn from talking to composing a communication product. Thinking of your audience for such products as a collaborator, a perspective you learned in Chapter 2, means building trust and acting with civility. But sometimes, often without meaning to, you do harm to others in your content and style. Avoid doing so by addressing the common issues discussed in this section.

Discriminatory or Derogatory Language

Without your awareness, bias against someone of a different gender, age, or ethnic background can surface in your writing. It's not difficult to avoid discriminatory or derogatory language if you think about it. Keep alert to changes in what groups of people call themselves or want to be called.

To avoid sexist language, for example, eliminate nouns and pronouns that indicate gender:

> *Sexist*: An average engineer spends 12 hours of his day at his computer.
>
> *Inclusive*: An average engineer spends 12 hours a day at a computer.

Or use the plural:

> *Sexist*: A scientist should be aware of the implications of his work as he decides what experiments to conduct.
>
> *Inclusive*: Scientists should be aware of the implications of their work as they decide what experiments to conduct.

Or recast the sentence:

> *Sexist*: Profile a client so that you can write your proposal to meet his needs.
> *Inclusive*: Profile the needs of your client; then write your proposal to match those needs.

It's fairly easy, then, to avoid calling attention to gender in nouns and pronouns. It's more difficult to avoid nouns that carry connotations derogatory to women, the elderly, or people of a different race or color from yours. The context for your remarks is critical. Sometimes, authors match terms in derogatory ways, as in a reference to the men and the girls at the office; men should match women; boys match girls. The main point is being sensitive and polite, attributes that would have prevented a court case, for example, in which a man claimed he was being harassed by a waitress who kept calling him "Honey."

In addition, avoid using yourself as a standard. For example, *Asian* is preferable to *Oriental* because the Orient is east only from a Eurocentric perspective. *European* is preferable to *Western*. Similarly, nonwhite implies a norm of "whiteness"; "flesh colored," meaning beige or pink, ignores flesh of other colors. A US publication calls its January issue the "Winter Edition"; that designation is valid in the United States, but not, for example, in Australia, where January is high summer. The author of an article about problems in the tech industry dated the events as "during the Thanksgiving weekend." Countries celebrate Thanksgiving at several different times (often related to harvests in their geographic region). Some countries don't have such a celebration.

The terms by which we talk about people in society change, sometimes rapidly, over time. The changes may seem to go overboard, a response sometimes called—usually in a negative sense—"political correctness." But as an ethical communicator, you're responsible for keeping up with the changes.

Misleading Technical Terms

Although technical terms are essential in technical communication (of course), such terms, appropriate for people who share your understanding, may violate the trust of other readers. Accurate, they may still mislead. To help the public understand scientific developments, you may need to use more common, less precise terms.

Increasing public comprehension is also a major ethical concern underlying the legal forms people encounter while conducting their everyday life, including consent agreements that accompany the use of software and websites. As contracts, these agreements, along with product warranties and purchase and refund statements on airline tickets, should adhere to the common law guideline that these documents are acknowledgments between parties who can understand the terms. But many contracts use language that disserves the less technical reader. The US federal government and many states have developed guidelines to encourage "plain language" in such contracts (see box).

PLAIN LANGUAGE

Government agencies as well as corporations and other groups are increasingly requiring that their guidelines and other documents conform to standards of plain English. This shift from highly technical or obscure language requires a new style of thinking and writing to deliver complex content. The goal is to help people understand a document's meaning without changing its substance.

Government officials hope that the use of such language will make their pronouncements more trustworthy. Marketing professionals hope the language will help their messages reach customers better in the context of the internet's information abundance. Companies hope plain English will increase workplace efficiency as employees understand corporate expectations better. Clearer messages also tend to evoke faster and more positive responses when companies write to clients and customers. Better-designed government forms reduce time for both citizens as they fill in the forms and employees as they process them. That reduction in time also offers financial benefits.

To counteract the "consent fatigue" that can occur when users of smart devices encounter "sets of long and unreadable data handling conditions," the IEEE (2019) calls for simplified data privacy warnings about how the devices will collect their personal data (p. 23). These guidelines aim to restore good faith and meet an ethical requirement on the part of the designer or provider of the product.

Deceptive Visuals

As one illustration of how visuals can both inform and deceive, consider maps. For years, maps have been used for multiple purposes, including advancing national ambitions. As one researcher argues:

> A good map tells a multitude of little white lies; it suppresses truth to help the user see what needs to be seen. Reality is three-dimensional, rich in detail, and far too factual to allow a complete yet uncluttered two-dimensional graphic scale model. Indeed, a map that did not generalize would be useless. But the value of a map depends on how well its generalized geometry and generalized content reflect a chosen aspect of reality.
>
> (Monomer, 1991, p. 25)

For more about using visuals effectively and ethically, see Chapter 9

Technical professionals have to ensure that any suppression of truth in a visual does not mislead the reader.

Lack of Disclosure

As a technical communicator, recognize and disclose any limits on your information. Discuss problems, wrong terms, failed prototypes, as well as successes. The pressure to exaggerate positive results, particularly in commercial endeavors, is strong. Some selectivity is natural, but your frank disclosure of mistakes could help other researchers avoid them or assess their significance. Don't be afraid to say, "We were unable to . . ." or "We had to reduce the number of samples because . . ." Never be afraid to admit when you're wrong. As the IEEE guidelines argue, creators and operators of A/IS have a strong ethical obligation to ensure transparency in explaining and interpreting their innovations to all impacted stakeholders and in clarifying the underlying rationale for their design.

Inappropriate Postings on Social Media

Social media are unparalleled in their ability to connect people and disseminate information. Their benefits, like the benefits of digital technology in general, are well acknowledged and real. You are likely to depend on social media now and in your career, both to network with friends and colleagues and to learn and create knowledge and understanding collaboratively. You'll learn strategies for such knowledge acquisition and interpretation in Chapter 6.

But the influence of social media is not always benign. It can be a source of disinformation: lies, distortions of fact, politically biased and slanted statements of opinion, and overwhelming numbers of automatically generated messages. It has a serious dark side. It can amplify messages in ways that reduce possibilities for social fairness and decency. Your instinct for ethical behavior in the global community must be always on as you contribute to and depend on social media.

Every day, as you write and as you talk with people in person, online, on the phone, or in a video conference, keep in mind the central value of ethical communication: respect those other people. To do so, in broad terms, demonstrate that you are trustworthy and empathetic; that you are acting in a way intended to foster the greatest good for the greatest number; and that your action is not only good, but *right* for the community.

CHECKLIST: COMMUNICATING ETHICALLY AND PROFESSIONALLY

1. Develop an "instinct for community" to govern your behavior in all situations
      Cultivate your good character to ensure your actions will be right for others
      Seek the greatest good for the greatest number
      Determine if you action is not only good but also *right*

2. Adhere to any explicit laws regarding your interactions with others
3. Follow applicable codes of practice
      Published by professional association and journals in your discipline
      Maintained by organizations where you work or study

4. Design ethical uses of technology
      Use digital technology to serve others
      Advocate for the environment and the rights of those who are underserved

5. Build trust and civility
    Engage in dialog to disagree without being disagreeable
    Avoid common ethical issues in content and style
        Discriminatory or derogatory language
        Misleading technical terms
        Misleading visuals
        Lack of disclosure
        Misleading postings on social media

EXERCISES

1. Review the syllabi or other statements distributed by three instructors of courses you have been enrolled in about protocols for conduct in class. How extensive are such guidelines? What attitude toward students do the guidelines reflect? Among other purposes, a syllabus establishes the authority and character of the teacher. Does the syllabus show an instructor who is friendly? Intimidating? How are those attributes demonstrated in the syllabus—for example, in word choice, choice of details, length of the syllabus, format, and design. What do the syllabi suggest concerning the level of trust among students and between students and the teacher in that class? Do they cover instructional situations both in person and online? Do they differ in what they suggest about the culture of different classes? Write a brief memo to your instructor about strategies employed in the syllabi to persuade students to comply with codes of classroom conduct.

2. In Chapter 3, you read sample passages from codes of conduct issued by professional associations and a corporation. Your college or university also most likely posts such statements on its website. These statements are becoming more common, and more detailed, in the complex and changing multicultural environment of the 21st century. Download or otherwise obtain a statement about ethical behavior from your university or other organization you work for or belong to. Then write a brief report analyzing, at least from your point of view, the persuasiveness of the statement and whether the organization lives up to stated goals.

3. Find a few examples of ambiguous or otherwise misleading language online or in social media. At your instructor's direction, circulate links to those examples to the class. In a cover note, explain why you think the person or organization posting these examples is either unconsciously providing misinformation or deliberately circulating disinformation—that is, lying.

4. Find a visual that could be construed as misleading. For example, you might choose a chart from Spurious Correlations (https://bit.ly/3xKxHtZ). Be prepared to discuss the ethical implications of presenting the visual as

fact or out of context. What harm could that deception do? Who might be affected? What is your responsibility as an ethical technical communicator? (Courtesy of Rebekah Smith)

FOR COLLABORATION

An excellent approach to showing respect for others is to engage in projects that enhance your community, understood as a local, regional, or larger community. You may already do so on campus or in your neighborhood. Seek out or create such an opportunity with a few teammates. Create a new project with others in your class, or enlist classmates in a project with which you are already involved, or be prepared to report to the class on your collaborative activities on a project outside of class, perhaps in a service-learning placement. At your instructor's direction, prepare a written or oral report (or both) on the project to be presented at the end of the term. During the term, use design-thinking activities to advance each phase of your work.

REFERENCES

AIA. (2020). *Framework for design excellence*. Retrieved from https://bit.ly/3tEpSnH

ASME. (2018). Society policy: Diversity and inclusion. P-15.11. Retrieved from https://bit.ly/3xcDdpt

Coghill, A., & Garson, L. (Eds.). (2006). Appendix 1–1. Ethical guidelines to publication of chemical research. *The ACS style guide: Effective communication of scientific information* (3rd ed.). Washington, D.C.: American Chemical Society.

Jones, N. (2016). The technical communicator as advocate: Integrating a social justice approach in technical communication. *Journal of Technical Writing and Communication, 46*(3), 342–361.

IEEE Global Initiative on Ethics of Autonomous and Intelligent Systems. (2019). *Ethically aligned design 1st edition: A vision for prioritizing human well-being with autonomous and intelligent systems*. Retrieved from https://bit.ly/3arUgtZ

Microsoft. (2020). *Achieve more. Standards of business conduct*. Retrieved from https://bit.ly/3sykGQV

Monmonier, M. (1991). *How to lie with maps*. Chicago, IL: University of Chicago Press.

Ross, D.G., Oppegaard, B., & Willerton, R. (2019). Principles of place: Developing a place-based ethic for discussing, debating, and anticipating technical communication concerns. *IEEE Transactions on Professional Communication, 62*(1), 4–26.

Shaw, M. (2020). This luxury tower has everything: Pools, a juice bar, and flood resilience. *New York Times*, April 29. Retrieved from www.nytimes.com/2020/04/29/opinion/climate-change-architecture-design.html

Wu, T. (2020). Bigger brother. Review of Shoshana Zuboff, *The age of surveillance: The fight for a human future at the new frontier of power*. The New York Review, April 9.

4 Communicating Collaboratively

"Working together, talking to each other, working in a more agile way. People are probably not so fixed any more in their working environment," argues a manager at an international bank. "They work much more in projects" (Bray, 2016, p. B6). In this chapter, you will learn strategies for working together—collaborating—in the projects that are at the heart of design thinking.

This chapter first reviews the attitudes and practices that enhance any collaboration, from conversations through the creation of communication products. The next section discusses activities through which collaborators can generate innovations. The chapter then turns from ideas to plans—that is, implementing and sustaining innovations, at scale, in the world. The discussion reviews strategies for composing, leading, and otherwise participating in a team, in both a physical and digital environment, locally and globally. The chapter ends with a look at two ways, among many, to compose a communication product collaboratively. One way is document-driven: Accepted genres and conventions provide a ready framework for assigning team member tasks and composing technical and scientific documents. The second is an every-one-composes-together approach. This approach aligns well with design thinking. A team of entrepreneurs may use such an approach, for example, to craft communication products that will launch their one-of-a-kind enterprise.

COMMUNICATING TO COLLABORATE

For more about listening ethically, see Chapter 3

Start to collaborate by listening. Continue by taking turns in a conversation. Keep invested in the process. Avoid distractions to focus on the moment, the scene, and others.

Listening

Listening is not simply the absence of talking, but it is an important activity in itself. One former British police officer who specialized in hostage negotiations defines listening as "the identification, selection, and interpretation of the key words that turn information into intelligence" (Bartleby, 2021, p. 51). As you listen, you analyze what the other is saying, you assess the framework and emphasis of the speaker, what they are trying to achieve, as well as their emotional state and their values. When you talk, you're not listening. Keeping quiet gives you an advantage in knowing something about them while you limit what you reveal about yourself.

DOI: 10.4324/9781003093763-5

Be patient. Let the speaker finish without interrupting or trying to finish the speaker's thoughts. Be open to new ideas and experiences. Be ready to learn. Temper your competitive instincts. Adopt a community perspective. Build on ideas together.

Taking Turns

The aim in a conversation is to build a relationship, to show respect for each other and encourage trust.

Asking Appropriate Questions

Conversations build momentum through comments, often through a pattern of questions and answers. Be careful not to absorb your listening time with thinking about your next question. Let others talk and then, perhaps, restate what you've heard to confirm that others are being understood. When you do ask a question, invite more than a yes or no answer ("What factors in the economic or cultural setting of Vancouver do you think favor our expansion there?" rather than "Do you think Vancouver will work?"). Watch for perhaps unintended negative slants to your questions, as in these:

> "Anyone familiar with this situation would agree with our decision, don't you think?"
> "Do you have anything at all to say about this?"

Accommodating to What Others Say

A simple technique for taking turns is to incorporate a key word or phrase from what the other person says into your own response:

A: Should we consider recreational users of our system or only professionals?
B: Do you have a sense of the size of the professional market?
A: I'd estimate that market as probably smaller than recreational.
B: So, we should probably include recreational users in the plan, right?
A: Right. Let's include both.

You devalue another person's ideas or story if, after she has spoken, you simply launch into your own story or ideas. Think beyond your own personal contribution. You can also cause ill will if you let yourself be distracted and don't keep up with what someone is saying, or if you decide before others have stopped speaking that what they are saying isn't worth listening to or is wrong.

Being Invested

It's easy to be distracted, especially in a group discussion, perhaps especially in a video conference. During such virtual meetings, it's tempting to slip into a news-feed or other program or to be drawn to an interesting background. That's why some experts think telephone conversations are more productive, at least for one-on-one discussions. Nonverbal signals are important, however, because they may reinforce, amplify, or contradict what's being said. You need to assess the situation, who is saying what, where and when, in a group. Watch carefully what people do as well as what they say. Don't hesitate to ask others for advice to make sure you are sending and receiving the right signals.

On the Ground

In a physical setting, sit or stand at a comfortable and safe distance from the others. Community or cultural norms and guidelines may help you determine that distance as well as a good reading of their body language (are they moving away? moving toward you?). Maintain a welcoming posture. The meaning of postures differs across cultures. For example, crossing your arms in front of your chest in the US may seem off-putting, but it's a more welcoming posture in Sweden. Eye contact, too, may be desirable—or not. Avoid signs that indicate boredom, such as looking at your phone or staring out a window.

In a Video Conference

During a video conference, nonverbal signals shift somewhat as the focus is largely on an individual's face and upper body. In addition, the physical setting may convey information about participants without their being aware of its disclosure. When the setting is a home environment, increasingly common as people work from home, the particular setting should be a matter of choice, not just an afterthought. Where you position your computer can reveal much more than you want to about your taste—in books, in furnishings, in art. Consider adopting one of the several generic backgrounds (such as a picture of your campus or of a city at night) available on most platforms.

For more about participating in a video conference, see Chapter 19

Across Cultures

Consider the following elements of interpersonal communication style as you conduct conversations across differences in cultures, whether local, regional, or global:

- *Show of emotion*: Would the other person expect you to show emotion?
- *Self-disclosure*: How clear is the distinction between public and private information? Especially in opening a conversation, will the other person expect you to ask about family and other private issues? Should you expect to be asked such questions?
- *Directness*: If you need to make a request, should you phrase it directly and upfront or postpone the request until you can soften it with a buffer of other matters?
- *Imposition*: Similarly, how comfortable are you—and is the other person—with your asking a favor? Is self-reliance more valued? Is asking a favor taken as losing face?
- *Names and titles*: Will the other person expect you to use their name with some frequency in the conversation ("As you suggest, William, this approach has a few disadvantages")? That's often seen as effective in a US context. Or will the other person find that annoying (perhaps with a negative feeling that the usage is typically American)?
- *Leave taking*: How extensive should your expression of goodbye be after a conversation or a meeting? Is merely exiting the room or disconnecting from

a video conference enough? Or is that rude? Should you make a specific statement about your discussion ("I think we accomplished a lot today, thanks to
your good insights. Thanks so much")?

COLLABORATING TO INNOVATE

To generate innovative solutions in situations that pose messy problems, consider
engaging in the six activities identified by researchers at the Stanford d.school and
described below, using the researchers' key words for each phase (Doorley & Witthoft, 2012, p. 47). These activities aren't sequential steps but iterative elements.
They help frame the early stages of discovering or inventing a creative product or
service, including, of course, a communication product.

Most important, making these activities productive depends on gathering
collaborators who represent different perspectives, experiences, expertise, and
backgrounds. This includes the people you're designing for. Your empathetic understanding of them grounds the process. You listen to their voices, either directly or
as you advocate for their interests, particularly those who have been underserved
in your community or who lack the same leverage you have in getting things done.

Saturate

The initial activity is immersing yourselves in the design situation. You and your collaborators share ideas, stories, and information. In a makerspace or other physical
setting that you can own for a while, you post, for example, photos of the people
your project aims to serve and their environment. If you've taped interviews with
them, listen to them (perhaps more than once). Review any videos of your observations on site.

For more about
makerspaces, see
Chapter 1

Synthesize

In the second activity, you "create clarity from complexity." You see through the
messiness and abundance of information, omitting details, rearranging, synthesiz-
ing. As you interact, you often generate ideas that no one would have thought of
alone. Moreover, in explaining an idea to someone else, you clarify your own thinking. Sometimes, a casual, even offhand remark provides the kernel of a new way
to see the situation, to align information, and, thus, to frame the problem. As the
cliché goes, it's picking up the stick at the other end.

Focus

Given that frame, in a third activity, you "narrow on a single topic." You make your
information and insights fit the framework.

Flare

The fourth activity, the opposite of focusing, is "going big with ideas." You ideate—
that is, you generate "tons of new concepts and options." You brainstorm, mind
map, or create a storyboard or wireframe sketch of an overall project (Figure 4.1).

For more about this
story map project,
see Chapters 7 and 8

The rough sketch in Figure 4.1 shows the researchers' early thinking about the
project. Each horizontal pair of panels corresponds to a story segment, including

4.1
Wireframe Sketch
of a Story Map
Source: Stephens
and Richards (2020)

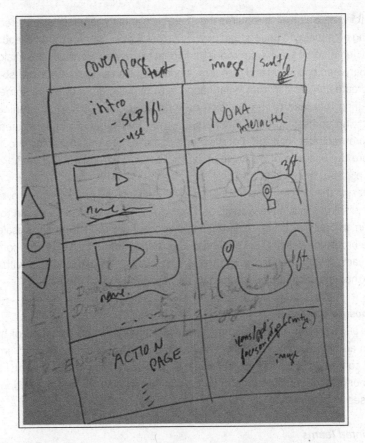

two introductory segments, maps and videos for each location, and a concluding page that orients users toward action. Text and videos are indicated on the left of each panel pair; photos or interactive maps are on the right.

Realize

The fifth activity in capturing your thinking is making your ideas *tangible*—that is, realizing them. A prototype draft, for example, is one such action when what you are producing is a communication product.

Reflect

The sixth activity is reflection. You collaborate to look back at what you've done. You learn from that experience what you need to do next.

The discussion you've just read looks like a list, a feature of its being printed on a page. But you go back and forth among these activities. The emphasis is on *activity*—not just talking, but doing, experimenting, trying things out, and learning from what went wrong—and right—to initiate another activity

TURNING IDEAS INTO PLANS

These six activities, based in design thinking, can launch innovative projects to solve problems and pursue opportunities. Carrying through on projects requires something else as well. It requires the strategic planning generally associated with working on a team. To understand how a design-thinking approach to collaborating may differ in some ways from more traditional teamwork, consider the analogy in style between *bricoleurs* and *planners*. A bricoleur is an artist or other craftsperson who creates an artifact from what happens to be lying around, trying something and letting that lead to the decision on what to do next, trying something and then trying something else. Planners, on the other hand, emphasize hierarchy and "premeditated control." For planners, mistakes are missteps; for bricoleurs, they are the essence of a navigation by mid-course corrections (Turkle & Papert, 1990, p. 136).

Design thinking tends to favor bricoleurs. But planning has to play a role, too, in accomplishing goals in the workplace. Planning is important in carrying out routine tasks in an organization or implementing a concept or product, perhaps scaled up from a prototype design. This section of the chapter offers strategies for composing a team, developing team identity and norms, and managing conflicts within the team. The next section discusses ways to manage the team process.

Composing the Team

Teams are either formed by someone outside them, such as someone of higher rank in an organization who is not on the team or a faculty member in an academic context, or they form themselves, an increasingly common activity, including among students. That difference is reflected in how the team functions, as you'll see.

Appointed Teams

One example of the first kind of team is a *standing* team, which serves for a limited term and attends to routine, recurring problems. Such teams often set policy in an organization or review actions to see if they comply with the policy. For example, your school may appoint a standing committee of students to advise the food service or to review matters of academic misconduct. Professional organizations appoint standing committees of members to regularly review their policies and procedures, such as their code of business conduct, to adjust to changing global environments.

In addition to standing teams, organizations also appoint ad hoc teams to address a particular one-of-a-kind problem. The teams disband when the problem is solved. In a class, the instructor may compose students into project teams with a well-defined goal aimed at advancing students' learning.

Self-Formed Teams

Increasingly, ad hoc teams—temporary, project-based groups of individuals—form themselves to pursue targeted projects, both within organizations and across organizations. In the networked environment for creative 21st-century work, independent entrepreneurs also gather subcontractors to pursue projects, collaborating and disbanding as opportunities present themselves. In an academic environment, too, you may collaborate on an idea that serves to draw you into a team with expertise to pursue that goal.

Developing Team Norms and Identity

Whether your team is appointed or self-formed, take time early on to put aside direct work as you resolve differences and pay attention to yourselves as a team. One approach is to gather the group in a "listening circle," either in person or virtually. Only one person can speak at a time, for a designated amount of time. There are no interruptions. Others listen. Participants are welcome to choose their own topics, including interpersonal issues and attitudes toward joining the project.

Norms for the Team's Collaborative Process

To develop norms—that is, guidelines for activities and behavior—a leader or facilitator might ask members to suggest, say, seven norms. Write these on a whiteboard or shared screen. The team then decides which to keep or delete. Norms include, for example, the appropriate length of a meeting, accommodation to individual schedules, topics for open discussion, and those for private conversations. Norms for appropriate behavior might include sanctions against private chats and online activities, such as shopping or checking email, that are hard for others to see but disrupt the meeting. More positive norms might explicitly encourage equity in turn-taking, especially for those who have less power or leverage in the larger organization.

Virtual meetings for people working from home can benefit from norms and rituals that serve to separate home and work. A ritual might be a stand-up meeting every morning at 9 to start the day, talk about goals, and solicit advice. Or, in combination with a norm of stopping work at 5, the team might gather online at that hour to talk about what they've done. The team might agree to a norm that includes dedicated time set aside each week for focused individual activity or leisure. Researchers have found that virtual team communication is best in bursts, for short periods, without allowing random interruptions. So that might be a norm.

A Common Team Identity

A common team identity, a sense of community, emerges from a baseline of shared trust, with more attention needed the greater the cultural differences and empowerment among team members. In discussing the successful acquisition of another company, an executive argues that culture is one of the most important considerations:

> Many transactions have failed for a lack of two cultures coming together. Both Celgene and Bristol Myers Squibb [the two companies] had a culture where you walk around the hallways and there's posters up of patients telling their stories. . . . We have multiple interactions, town halls, where we have patients come in and share their stories. . . . So that very much keeps both companies motivated, and that's where the cultures come together.
>
> (Elkins, 2020)

Such "coming together" can be particularly challenging, if essential, for teams that operate only virtually, including global virtual teams.

Without overemphasizing differences, the team needs to surface differences in practices and expectations, especially in the diverse teams at the center of today's knowledge economy. These differences range from the relatively superficial (such as accommodating every member's local holidays, time zones, and work hours in the schedule) to deeper differences in local policies, ideologies, and practices. People from different cultures see things differently. They often have different approaches to solving problems, setting goals, dealing with (or not dealing with) conflicts, apportioning tasks, and the like. They may also have different expectations about what communication products should look like and which forms apply to different situations.

Negotiating Conflicts

Surfacing and accommodating differences are essential elements of teamwork. In its policy statement on diversity and inclusion, the American Society of Mechanical Engineers suggests that managing diversity is the ability

to inspire and enable all people to

- align to a common vision
- communicate effectively and assure understanding
- know and accept what is of value to others
- leverage the strengths of others and trust their commitment to deliver as agreed
- appropriately recognize and celebrate successes often.

(ASME, 2018)

Here are a few strategies for negotiating the conflicts that may arise as you pursue these goals.

- *Focus on ideas and information, not personalities.* Emotional responses to individuals impede conflict management. Try to identify the source of the conflict in the *content* of the message rather than its communicator.
- *Pay attention to how you are perceived by others.* Nonverbal clues not apparent to you, such as books in your bookcase or photographs on the wall behind you as you participate in a video conference, may tell a story about you that contradicts what you say and may offend others or at least distract them from listening.
- *Agree on what you agree on.* Find common ground. Narrow areas of disagreement to specific ideas that can be addressed in the context of larger areas of agreement. Identify clearly the source of disagreement so you can focus on resolving it and not waste time rehashing what you already agree on. Common ground provides a place where mutual gains can be achieved.
- *Deal with what matters.* Weigh the significance of an area of disagreement and calibrate the energy and emotion involved in dealing with it to its importance. Don't get hung up on details or waste time over small points.
- *Discuss tradeoffs openly.* Don't pretend that "everyone wins" when major disagreements still need to be resolved. Specify what is being given up and what is being gained as the group moves toward consensus.

- *Accept what you can't change.* You can't use negotiation to change basic personalities or make people into something they're not. Accept differences. Sometimes, you can't change a person's mind.
- *Use objective criteria.* State problems so that you can use measurable standards. If you define success as "being the best," you increase the likelihood of disagreements. Instead, use an objective standard (e.g., "50 percent of the aggregate market") that can be the basis of rational discussion and ultimate resolution. Quibbling over qualitative words can engender unnecessary disagreement.
- *Move on.* Once a consensus has been reached, don't keep wrestling with the issue, raise it for further consideration, or try to modify the outcome.

MANAGING THE COLLABORATIVE PROCESS

A snapshot of design thinking would show people moving around, engaging in activities. For more traditional planning, it might show people in a meeting, either standing or sitting. The term *meeting* evokes a much-lamented feature of traditional teamwork. But you need both the activities of free-form ideating and prototyping as well as predetermined meetings to get your ideas to work. Here's a four-step overview of how to plan and manage a collaborative project. The discussion also suggests current thinking about making this process more agile and oriented to resources rather than tasks. More, that is, like design thinking:

- Scope the work.
- Schedule the work.
- Stay in touch.
- Satisfy all identified needs.

Scope the Work

Once you're on a team, your first activity with your teammates is to scope the work. On an appointed team, the person appointing the team bears primary responsibility for clarifying its goals, leadership, and procedures, including deadlines and deliverables. Making the assignment can be a complex communication endeavor, as any student who has had to respond to a class assignment can attest. What seems unambiguous to the instructor may evoke only confusion in those who need to fulfill the request. But through dialog with the person setting the team's direction, your team confirms what it needs to do, when, and can follow through.

Small teams can usually operate collaboratively without a leader. But larger teams need someone who assumes final responsibility for keeping people motivated and on track, helping to negotiate disagreements and conflicts, and consulting with stakeholders. That person may be

- Appointed by the person requesting the work
- Assumed by the person forming the team
- Elected by the team
- Acknowledged as the technical expert by the group

- Known as the gatekeeper for important information
- More committed than others on the team to seeing the project accomplished.

Self-formed teams are becoming prominent in the creative economy as independent professionals pursue projects with an ever-changing network of subcontractors (Spinuzzi, 2015). Design thinking helps to define the project. That definition of a problem or opportunity leads to a process for developing resources, including the right people. The ad hoc team combines and then disbands when the project is complete. Project-driven, self-formed teams are also becoming a feature of corporations seeking innovative and swift responses to a rapidly changing global market. Students, too, are forming teams to work on projects they initiate, often alongside faculty but not in response to an explicit, predetermined faculty assignment.

On such project-driven teams, you develop a pattern of talking about what you need to talk about, challenging each other's assumptions as you question and learn from each other. You clarify what success will mean as you negotiate the differing levels of interest each of you may bring to the project. Assessing candidly who is on the team will help spot holes in expertise or experience that need to be filled. There is never a perfect team, but, if some critical skills are obviously missing, you have to remedy the problem.

Schedule the Work

Once the scope is confirmed and agreed to by all members, the next step is to divide needed activities into discrete phases and place them on a schedule from project launch to date for final completion. The schedule drives the workflow in a traditional approach, sometimes labeled a *waterfall* process. Things happen, of course, and may cause delays or other changes. But the "premeditated" plan coordinates team efforts and ensures accountability. Such a plan is particularly important when team members carry out their tasks separately. The division of those tasks has to be fair and perceived as fair, well understood, and without overlap of assignments given to two different people.

In a bricoleur-like approach oriented to design thinking, teams can operate in a more fully collaborative way. A popular workflow style, for example, called *agile*, encourages teams to schedule the work in shorter time frames of a few days (one technique calls these "sprints") in which members commit to finishing a subset of activities toward the larger project goal. Collaborative reviews and retrospectives at the end of each sprint set procedures for the next. The approach is less about individual tasks, individually performed and later blended, than about all team members as resources for each other in pursuing a joint goal. The plan emerges in each sprint and is revised in the next sprint, rather than being established at the outset.

Stay in Touch

In a schedule-driven approach, knowing their assignments, individual team members settle into their tasks and may regard checking in with the team or the team leader as a distraction. Members work on their own over even fairly long periods. Without regular meetings, however, team members may lose sight of the overall objectives. Formal team meetings remind members of where they're headed and the role of each member in getting there. The meetings help ensure accountability.

An agile approach depends on continuous communication. As they talk about the problem and potential solutions with each other and other stakeholders, including clients and customers, team members discuss options and procedures openly. Frequently, sometimes daily, the team gathers to update each other (in what are sometimes called "scrums") on what they have done, what they're doing now, what is going well and not so well. They get advice and hold each other accountable. At the end of each short time frame, they review and reflect before plotting their approach to the next period. A manager at a large technology firm calls this process "the loop: observe, reflect, make."

Choosing the Team's Technology

Staying in touch requires a decision on the technology appropriate to the team's project. Workplace collaboration software such as Google Workspace, Slack, or Microsoft Teams offers multiple ways to communicate through video and text messages. The software also allows the team to archive and manage content digitally, both when members are co-located and when they are distributed across regions. The choice of technology may be dictated by the organization where the team is employed or studying. Or the team may choose its technology as it sets its norms. Given a choice, consider using the simplest technology that will do the job—knowing, of course, it could be the most advanced. Avoid a technology choice that might

- Intimidate potential team members
- Represent a form of technological imperialism to those less technologically advanced
- Cost too much, given the team's budget
- Take some team members more learning time than they're willing to spend
- Disadvantage members whose differences in abilities do not allow them access.

Meeting

Staying in touch also means meeting. This section focuses on more formal meeting practices rather than those associated with agile and flexible approaches. As a collaborator, you need to understand the meeting process. In meetings, you advance organizational and personal goals and engender trust while demonstrating interpersonal and technical competence.

WHY MEET?

Meetings are useful to:

- Build trust in a team that will later work mainly remotely
- Monitor progress on a project that has run into serious problems
- Gain buy-in for a concept or new procedure
- Foster consensus on a controversial issue
- Launch with fanfare a new product or initiative.

All meetings take time, and many people, students and professionals alike, think meetings mostly *waste* time. To avoid such waste and be productive, if you are in charge, set an agenda and follow it, as in the box on "Leading a Meeting by Following an Agenda." Begin the discussion on time; this advice, however, may have to bend to local customs. One corporate technical writer comments:

> I worked with a team of Brazilians (they were the client), and they were habitu-
> ally late for every meeting. They also didn't stick to the agenda. If someone from
> their team had an ad hoc idea, everyone was expected to work on that idea. Our
> meeting was, say, nominally two hours long, we would frequently go over the time
> period and just work until everything was covered. It was like a meandering river.
> They just used time differently. The project was successful, but that was because
> we Americans recognized that they were the client and we had to adjust to their
> norms.
>
> (Luttrell, 2020)

LEADING A MEETING BY FOLLOWING AN AGENDA

Before the Meeting

In announcing the meeting, confirm the time (noting time zone for a virtual event) and place (if face to face); provide the agenda; and attach any documents, such as the minutes of an earlier meeting or details of the matter to be discussed that attendees should read in advance.

To Start the Meeting

State the purpose briefly but specifically, both the process to follow and desired outcome. If there's bad news that needs to be addressed, start with it up front if you think attendees won't pay attention until the news is out.

> *Example*: "After a brief update from Sue and Roger about their research on alternative routes for the bypass as you read in their report sent earlier, we'll open the meeting to general discussion of the merits of each and establish a priority list for further testing."

During the Meeting

Arrange items for discussion strategically. In general, quickly cover old business first (an attachment to the announcement may be enough to provide background and save time in the meeting). Then, introduce topics you antici-pate will engender only brief discussion before those that need more exten-sive discussion, the noncontroversial before the controversial.

As the discussion develops, either you or someone you appoint should take notes, formally called minutes, to record the substance emerging from the time spent: a summary of key information presented, decisions taken, and actions, including noting who has been made responsible for what, to follow up (see Chapter 15 for more advice on minutes).

At the End

Summarize briefly, confirm future steps and tasks (and the assignments for carrying them out), and perhaps set a date for the next meeting.

As Luttrell observes, meetings assume a rhythm that can vary across cultures. One research study, for example, found that Chinese participants are more comfortable with silence in a meeting than participants from the US. They are more likely to allow a discussion to unfold gradually and circuitously, returning to points that US participants had thought settled. US participants tend to value speed, getting things done. The study used an analogy to describe such differences, noting that the links "needed to complete a task were like the vertebrae of a large, flexible dragon rather than a rigid-backed steer" (O'Hara-Devereaux & Johansen, 1994, p. 211).

Internationally, English is often used as the language for discussion as many professionals, whatever their native language, are also fluent in English. But not always. The box on "Speaking and Listening across Languages in a Meeting" provides advice for participating in a meeting where language is an issue.

SPEAKING AND LISTENING ACROSS LANGUAGES IN A MEETING

- Be especially patient: discussions where language itself becomes an issue tend to take longer and be more tiring.
- In small groups and meetings, avoid side conversations in another language because these may reduce group trust and engender suspicion.
- Focus on content: don't correct a speaker's pronunciation or grammar.
- Don't rush in to supply what you think is a word or phrase eluding the speaker.
- Speak carefully, but avoid the loud and monotone delivery some Americans consider the essence of communicating across cultures ("THE RECOVERY PLAN REQUIRES 5 STEPS").
- Ask for clarification, politely, if you can't understand what someone has said.
- Streamline your vocabulary and use formal English, avoiding slang.
- Avoid jokes and diversions.
- If an interpreter is aiding the interaction, talk with the other participants, not the interpreter.

Satisfy All Identified Needs

Whatever workflow style the team adopts, from a premeditated plan to more agile practices, the final phase should include reflection not only on what the team accomplished, its product, but also on its process. You learn to be an effective team member by working on effective teams. Making explicit the nature of the team's processes improves the chances for individual and team learning. When the project is complete, celebrate—and then discuss what worked and what didn't. How can strong features of the process be carried into other team situations? How can weak ones be avoided in the future? What individual member's skills need to be refined or augmented? The team should ask what it can learn from its experience.

CREATING COMMUNICATION PRODUCTS COLLABORATIVELY

Communication activities integrate with technical ones as you scope, schedule, stay in touch, and satisfy identified needs. You listen and talk as you work. You also compose communication products, creating text and visuals, in multiple media, to present the result of your work to the wide variety of audiences who have an interest in it. Your communications emerge from and track the work.

For more about arranging content to meet audience needs, see Chapter 2

Document-Driven Plan

When you know that your project aims, for example, toward a final document, either because it is a routine problem-solving task—for example, an update on a policy—or an assignment from a client or professor who has requested a report or proposal based on your investigation, then that genre of document can drive how you scope and schedule the team's work, including composing and testing a prototype draft and delivering the final document.

Agree on Your Purpose

First, agree on your team's purpose for writing the document and your readers' purpose and needs in reading. Reduce your assignment to the most concrete terms possible as you plan the requested final deliverable. Clarity at the beginning is essential. For example, the two sides in a multinational team designing a joint venture in the chemical industry had different expectations for the business plan they were writing. The Americans thought they were creating a sketchy "vision" document, long on speculation and short on financial information. However, their North European colleagues thought that even an early document in their negotiations should present detailed financial information. When they couldn't agree on the genre, the team recognized even deeper incongruities, which caused them to abandon the venture.

To avoid working at cross purposes, write a statement of purpose about the document that all can agree on. Here's one, from the student team competing for start-up funding you read about in Chapter 2. The resulting business plan appears in Chapter 14.

· Purpose: To write a business plan describing our *QuadCrew Rowing System* that effectively and persuasively meets the specifications of the University's HenHatch competition for start-up funding.

Expand your purpose statement to develop a preliminary outline that forecasts the structure of the document as a whole. Agreeing on an outline helps the team negotiate conflicting opinions, evidence, and formats before positions harden into content—that is, into text and visuals. In addition, the outline lets an instructor, client, supervisor, or sponsor preview the work to see if it will meet requirements and correct the course of the project early on.

Assign the Writing

Once your team has settled on a purpose, genre, and preliminary outline, that track gives you a variety of options in assigning discrete roles to team members. One is to have each team member research and write one segment of the report, with perhaps one member taking on the added duty of editor in the last stages of the project. Another is to divide responsibilities by team member skills, expertise, and interest in the project, with one team member becoming the primary writer. The designated writer pulls together information from meetings and from drafts submitted by team members as they complete stages of their work and prepares a draft final document, which is posted in the team's collaboration software or content management system for comments. The writer revises the draft to incorporate the comments, edits it, and prepares it for delivery to the audience.

Whether organized by segments of the document or by roles that include a designated writer, a document-directed plan engages all team members in producing text themselves as they go along. It's not up to the writer to make sense of the entire team effort.

Develop a Schedule

For more about designing schedules, see Chapter 9

To monitor the completion of each task and progress toward the goal, develop a schedule, with names of those responsible clearly indicated, along with due dates for content to be submitted, probably to a content management system all can access.

Social Writing

The document-driven approach governs what can be called a cooperative endeavor. Team members work largely independently after an initial launch and before a final blending of efforts. It works well on teams of any size, even large ones, when the product to be delivered is in a relatively standard genre.

Another approach is what a student practicing such an approach in an academic makerspace calls *social writing*. It emerges from more agile practices in collaboration and depends on a well-coordinated network of members. Everyone composes together. Robust digital technology makes such collaboration easy, whether the team is co-located or working across distances. It's an open process of writing quickly and looking at each other's writing throughout the work.

Reflecting on their work at intervals throughout the project, team members reconsider their experiences and evaluate past activities as they impact what's to come, all the while turning knowledge they've gained into text and visuals. That process may take advantage of a shared content management system in which the team's text and visuals are stored in modules. These are not pegged to sections of a document but to different topics derived from the project itself.

These self-standing units of information can be assembled in different structures for different audiences, contexts, and purposes as the project develops. Such an approach is particularly effective for networked teams whose members may come and go as they take on projects that address multiple audiences and can profit from a database of texts to be mixed and matched for each new situation.

Social writing can work as a strategy for producing a traditional document. But especially in situations where an innovative communication product is needed, it offers the promise of more creative crafting and delivery of such a product. Social writing in that sense is less about writing than about meeting the communication needs of a larger community.

One Voice

Regardless of how a team produces a final communication product, the product has to speak with one voice. It has a consistent persona. That may represent an organization's voice, enhanced by a staff of communications professionals. Or it may be the persona of the lead writer on the team, agreed to by team members. Content also has to be consistent. No part should contradict or call into question another part. Introductory and concluding sections should mirror each other, and evidence presented in the middle should support both.

For more about ensuring consistency, see Chapter 11

Develop a Style Guide

Developing a style guide helps you reach concrete decisions about the form and expression in the final product. Refer to standard sources, such as any guidelines published by your academic department, organization, professional publication, or professional association. Even a brief list of stylistic preferences will help save time during the final editing and will forestall potential disagreements as team members draft their contributions. Here are some topics you might cover:

- Length of the document as a whole and of segments
- Format for introductory paragraphs to each section
- Method for citing sources
- Number and type of graphics
- Glossary of special terms and their meanings
- Format of lists and headings
- Punctuation preferences.

Negotiating matters of style is particularly important when you write on a multinational team. Separate negotiations about style from negotiations about content. Don't let a debate about appropriate use of commas, for example, sap energy and goodwill that should be reserved for issues of policy or scientific understanding. Establish a list of major terms and their equivalent in the target language for any document that will be translated.

DOCUMENT PLANNING WORKSHEET

Title of document
Today's date
Date document is due
Who is the document for?
    Primary reader
    Intermediate reader (if any)
    Secondary readers
What is the purpose of the document?
    Definition of project problem or opportunity
    Statement of purpose for the project
    Statement of purpose for documenting the project
What will the document cover?
    Main point
    Major subpoints
    Plan of sections (attach current outline)
Who needs to review the document for technical accuracy and completeness?
    Who needs to review for style?
    What will the document look like?
    Estimated length
    Page size and image area on page or screen
    Production method (print or online)
    Typeface
    Images
How will the drafts of the document be distributed to team members for comments?
What is the document's timetable? (Attach detailed schedule.)
Contact information for all team members.

Record Your Approach

A good control system is important when you're working individually. It's essential when working collaboratively. The Document Planning Worksheet shows a fill-in-the-blank form for recording the approach in a team writing project. A schedule and document outline would be attached.

This chapter has reviewed strategies for working together—collaborating—on the projects that are at the heart of design thinking. Basic to any collaboration is respecting others, listening and taking turns in a conversation, and being invested in the scene, the moment, and the project. Through the activities of design thinking, generate ideas, clarify complexities by narrowing and focusing, create a prototype, and reflect, all iteratively. Then develop a plan to implement those ideas at scale, in the world. Think in terms of bricoleurs and planners as you choose your approach. Manage a team process by scoping the work, scheduling the work, staying in touch, and satisfying all needs. To compose a final communication product that will deliver your work to your audience, you might use a document-driven strategy. Accepted genres and conventions provide a ready framework for assigning team member tasks and composing technical and scientific documents. The second approach is an everyone-composes-together one. A team of entrepreneurs may use such an approach, for example, to craft communication products that will launch their one-of-a-kind enterprise.

CHECKLIST: COMMUNICATING COLLABORATIVELY

1. Communicate to collaborate
    Respect others
    Listen
    Take turns in a conversation
    Be invested

2. Collaborate to innovate
    Immerse yourself in the problem or situation
    Create clarity from complexity
    Narrow to focus on a single topic
    Flare, that is, generate multiple, even wild, options
    Make thinking tangible in a prototype
    Reflect, reconsider, transition to next activities

3. Turn ideas into plans and strategies for teamwork
    Compose the team, either appointed or self-formed
    Develop norms for the team's collaborative process and a common team
        identity
    Negotiate conflicts through focus on goals and content, not personalities

4. Manage the collaborative process
    Scope the work
        Clarify and agree on the assignment, if an appointed team
        Through iterations in an agile approach, flare, then narrow on an
            agreed goal

    Schedule the work
        Let a schedule set at the beginning drive traditional teamwork
        Use shorter time frames to let a plan emerge and change in an agile
            approach

Stay in touch
> Use meetings effectively to advance the team's work and enhance accountability
>
> Choose the team's appropriate technology
>
> Structure meetings with an agenda to enhance productivity

Satisfy all identified needs
> Reflect and iterate through your process
>
> Be a learner, review what worked and what didn't
>
> Celebrate

5. Create the final communication product
    > Let the specific document genre needed drive your work
    >
    > Collaborate to innovate a new format or revised genre for your community
    >
    > Ensure that the product is consistent in its persona and content

---

EXERCISES

1. Write a brief description of a group to which you belong (a sports team, club, professional society, or class project team). In your description, note how the group was composed and how it operates: how members are added, how they communicate with one another through formal and informal channels, how meetings are conducted. Comment, too, on the culture of the group in the terms you learned in Chapter I.
2. In any team project, a major activity is choosing the team's information and communication technology (ICT). The options expand daily. Select one form of workplace collaboration software, try it out, if possible (even for-profit corporations usually allow free downloads for limited times), and read about it online. Then write a brief report about its potential advantages and disadvantages for a collaborative project you are working on or one you might imagine working on in the future.
3. If you are working on a team project, develop a statement of norms for your activities as you read about such a statement in the chapter.

FOR COLLABORATION

The best collaborative exercise for implementing the advice in this chapter is to participate in a collaborative project aimed at solving a problem or finding an opportunity. Create a statement about the purpose and deliverables of your project. Record your team's approach to scoping and scheduling the work, staying in touch, and satisfying all needs. Complete the document planning worksheet shown above or adjust that worksheet to fit your project's specific approach. Keep aware about not only the product you aim to produce, but also your process for doing so, and, in a final reflection, see what worked and what didn't.

REFERENCES

ASME. (2018). Society policy: Diversity and inclusion. Retrieved from https://bit.ly/3xcDdpt

Bartleby. (2021). Hear, hear: The secrets of successful listening. *The Economist*, January 23, 51.

Bray, C. (2016). Your desk? Take your pick. UBS rethinks work space in its London offices. *The New York Times*, November 4, B6.

Doorley, S., & Witthoft, S. (2012). *Make space: How to set the stage for creative collaboration*. Hoboken, NJ: John Wiley.

Elkins, D. (2020). Collaboration is the best medicine. *UDaily*. Retrieved from https://bit.ly/2QhwZE4

Luttrell, C. (2020). Personal communication.

O'Hara-Devereaux, M., & Johansen, R. (1994). *Globalwork: Bridging distance, culture, and time*. San Francisco: Jossey-Bass.

Spinuzzi, C. (2015). *All edge: Inside the new workplace networks*. Chicago, IL: University of Chicago Press.

Stephens, S.H., & Richards, D.P. (2020). Story mapping and sea level rise: Listening to global risks at street level. *Communication Design Quarterly*, *8*(1), 5–18. Retrieved from https://bit.ly/3arXxt4

Turkle, S., & Papert, S. (1990). From hard drive to software: Gender, computers, and differences. *Signs*, *16*(1), 1–202.

Part 2

Managing Projects through Design

5 Defining the Problem or Opportunity

As a student and as a technical professional, you will spend a good deal of time dealing with all kinds of problems. One approach is simply to ignore them, although that may have consequences down the road. Another is to decide that intervening now is unnecessary or would be counterproductive. An example might be a doctor's taking a "watchful waiting" or "time will heal" approach to a medical issue presented by a patient. The symptoms may not indicate an underlying problem.

This chapter focuses, however, on situations in which problems need more active solutions. At the simplest level, the solution is known. You carry out known tasks to solve a problem. Those tasks may be assigned by someone else—for example, an instructor or supervisor. Or you or your team may implement them on your own. You learn and then apply the skills needed to make completing such tasks a routine matter.

Solving routine problems remains a large part of daily life. But the global 21st-century environment of knowledge work poses challenging and complex problems that go beyond the routine. Through design thinking, you observe and empathize with others as they interact with the environment. You seek ways to make that interaction better—for them, for you and your enterprise in the global marketplace, and for the environment, broadly understood. You think beyond preconceived ideas of what's needed in a situation to innovate. You find opportunities for new products and services where others didn't look, even where those who stand to benefit from your innovation may not have realized they needed it. Emerging technologies play a major role in the shift from thinking about problems to thinking about opportunities. You move from conventional approaches—that is, implementing the ideas you think *with*—to innovative approaches. You generate new ideas, the ideas that you think *of*. This chapter begins a series of three chapters in which you'll learn strategies for launching and managing a project to solve problems and investigate opportunities. The process begins with defining the problem or need.

LAUNCHING A PROJECT

The young researcher in her age-simulation suit you met in Chapter 1, with her colleagues, initiated the kind of project that is the focus of this chapter. It all begins with being curious about and understanding what someone or a group of people might see as a problem and what is needed to remediate it. In the process, you

DOI: 10.4324/9781003093763-7

Known problem ―――――――― **Known solution**	Known problem ―――――――― *Unknown solution*
Unknown problem ―――――――― **Known solution**	*Unknown problem* ―――――――― *Unknown solution*

"filter information, remix modes, and remake practices" to solve problems in new, multimodal ways (Dusenberry et al., 2015, p. 209).

To start the process, think of four situations for defining a problem or opportunity, as shown in Figure 5.1.

Known Problem—Known Solution

The key question, of course, is "Who knows?" Let's start with the first situation: known problem—known solution. This is relatively easy. For example, you need a new printer for your computer. You know the problem. Various manufacturers of computer printers offer a variety of solutions. They offer products to solve your problem. You match your specific need to the right solution at the right cost for you and buy one. Done. Such situations can become more complex when enterprises buy products and services from vendors, when your university, for example, buys computers from Apple. But the relationship is still the same. Your university knows it has a need. Apple knows how to solve it and would be pleased to do so. And so on. Much of the world's business fits this situation. Conventional thinking, both technically and in communication approaches, generally works.

Known Problem—Unknown Solution

The second situation introduces a new variable. Although the problem is known, the solution is not known. Let's say a company developing a new financial app has nine employees but expects to expand soon. The three founders, college friends in Houston, have worked from home or in a coffee shop. They coordinated their work through Google Docs, which they had used in college. But as their project took off, they added a network of others to their team. They also enhanced their client base, and those new clients were more sophisticated users. The team needed more robust technology for messaging, structuring and managing content, and video conferencing. New workplace collaboration platforms, including one from Google, entered the market, and they had to figure out the right one for their growing international team.

Here's another example of finding a new solution to a known problem. Makerspaces on college campuses offer attractive environments for teams to work on design projects, but they can be crowded and noisy. Complaints about those conditions have become a source of problems to be solved and have launched new projects to solve them. Let's say that, over several days in your department's makerspace, you observe the highly uneven distribution of people working there. Your team takes on the project of flattening the curve—that is, spreading the demand more evenly. You try a sign-up sheet on-site, but for various reasons, especially no-shows and people who came anyway, it didn't work. Your team develops prototypes of a variety of smartphone apps that would provide real-time information about crowding and the availability of resources to alert users remotely about timing their work there.

Unknown Problem—Known Solution

The third situation suggested in Figure 5.1—a known solution to an unknown problem—may seem puzzling. Can you have already solved an *unknown* problem? You can, in a very real sense. Someone once said that the secret source of problems is solutions. An emerging technology, developed to solve a problem in one situation, can meet a need that potential users in another situation cannot even imagine or articulate. A quick look at the offerings of an app store will provide abundant examples of apps adapted to meet previously unimagined needs.

Unknown Problem—Unknown Solution

Adapting apps for new markets moves projects in a more innovative and creative direction. The fourth situation described in Figure 5.1—an unknown problem and an unknown solution—is the most visionary. This situation provides the ultimate challenge for design thinking. You go wild and radical, ignoring constraints to develop new ideas, to create something entirely new. These are the life-changing innovations that entrepreneurs and corporations are after as they seek new markets.

CATEGORIZING THE PROBLEM

The previous section presented a grand picture of problem solving. This section provides a more detailed strategy for defining specific problems within three categories that are useful in technical, scientific, and professional settings. These are problems of fact, problems of means, and problems of value.

Problems of Fact

Problems of fact are at the center of basic scientific research. Researchers ask the fundamental question: *What is?* They observe phenomena and suggest explanations of how, and sometimes why, things occur. Physicists engaged in identifying stars, naturalists examining the feeding habits of quail, medical researchers looking for the causes of cancer or dementia or autism, epidemiologists tracing the origin of a virus—all deal with problems of fact. Such problems are driving current efforts at developing what is being called *big data*—extensive collections of statistics and surveillance analytics. Enabled by embedded electronics, global positioning devices, and other emerging technology, researchers in widely diverse disciplines and tech companies are assembling vast troves of data about people and phenomena.

Problems of Means

The dean of an engineering college likes to cite a maxim attributed to a Hungarian aerospace engineer: "Scientists discover the world as it is; engineers create the world that never has been" (Penn Engineering's New Dean, 2015, p. 26). Creating and building are essential to engineers and other designers. Their clients or the users of their products ask *how to* questions. Some such problems are routine. A problem solver knows how to do something and makes that solution known to the client or audience by building what was needed or answering a question. A known solution may still have to be adjusted to a specific environment or may have to incorporate new facts, such as a revised understanding about how a virus is transmitted, uncovered in ongoing basic research.

In addition to answering questions about how to do something, someone facing a problem of means may also have to answer questions about why something happened: *How come?* Why did an engineered structure fail? Bridges and highway overpasses fall, and planes crash (fortunately, not too frequently). Someone has to assess the circumstances of the failure, determine the possible causes, select a probable cause, and recommend what to do to get the item back in action—and to prevent such failures in the future.

Conducting research on problems of fact often requires solving a problem of means to investigate those facts. To investigate communication practices in a workplace, for example, researchers developed easily worn sensors, which they called "sociometric badges." These badges track how people talk to one another, who talks with whom, how people move around the office, and where they spend time, although not what they were talking about (Waber et al., 2014). The researchers provided such badges to thousands of workers in a range of offices who wore them on a voluntary basis. Their findings supported a generally accepted theory that knowledge is socially created, especially though random encounters. Based on such data, designers created the floorplans of workplaces to achieve a desired density of interactions considered appropriate for productive communication.

Problems of means can take on global significance. This is evident in the mission statements of many tech companies, research labs, governments, and non-profit, non-governmental organizations (NGOs) today. Many such enterprises say they want to make the world a better place. That's a big problem of means. These problems inspire the question: *What if?* Thinking about that question fosters innovation. An MIT researcher, describing a new device he and colleagues developed, goes big with such thinking:

> What if we were able to embed electronics in absolutely everything? What if we did energy harvesting from solar cells inside highways, and had strain sensors embedded in tunnels and bridges to monitor the concrete? What if we could look outside and get the weather forecast in the window? Or bring electronics to my jacket to monitor my health?
>
> (Zeeberg, 2020, p. D1)

Problems of Value

Should we allow sensors in our jackets to monitor our health? How necessary, in broad social terms, is getting weather forecasts on our windows or letting internet-connected devices reap data from our refrigerators? Problems in this third category—that is, problems of value—lead to an evaluation. You answer the question: *Should we?* To do so, you have to rate something, or rank priorities, or determine policy.

Some problems of value, as with some problems of means, depend on research targeted at problems of fact. In other words, you may need to measure whether something meets predetermined, objective standards. That's a matter of fact. The National Bureau of Standards and the Food and Drug Administration in the US, for example, focus on the development and application of standards to maintain levels of safety in a wide range of products and services. Standards worldwide that set, for example, the dimensions of shipping containers or require emission-control devices on cars help ensure the smooth running of the global economy. Standards may differ among nations, however, and thus require negotiation when multinational companies operate in different settings.

Problems of value require individuals and enterprises to assess priorities, determine trade-offs, and deal with the complex political and ethical issues knowledge workers face in the 21st century. Who decides what, in fact, makes the world a better place? And for whom? We can develop, for example, extensive data—facts—about changes in the planet's weather over the last many years. We can use sophisticated instrumentation to determine a range of means or methods for mediating those effects, including alternative devices for generating energy beyond coal-fired plants. But which is best? Which takes into account the wide diversity among the world's nations responsible for contributing to the change? How serious is this problem of climate change—if, indeed, it is a problem? These are problems of value.

For more about ethical challenges, see Chapter 3

ASKING DISCOVERY QUESTIONS

To home in on the definition of the problem at the core of your project, assign it to a category: problem of fact (*What is?*), problem of means (*How to? How come?*), or problem of value (*Should we?*).

That framework will help you engage in activities to generate options for its solution. Answering some or all of the questions in the box on "Discovery Questions

For more about such activities, see Chapter 4

and Answers: Processing Shipments at a Sporting Goods Store" may help inspire you (Broadhead &Wright, 1985–86). As an additional benefit, this sequence of questions advances your thinking toward modules of content to deliver in a final communication product for your audience.

DISCOVERY QUESTIONS AND ANSWERS: PROCESSING SHIPMENTS AT A SPORTING GOODS STORE

- *What is the situation and the background?* What information would an audience need to understand or be convinced that there is a problem? At the sporting goods store where you work, shipments are processed only haphazardly. Your manager has asked you to suggest a plan for better processing.

- *What is the problem itself?* No one has been assigned or has assumed responsibility for unpacking the boxes.

- *What are the effects of the problem?* Boxes remain unopened in the stockroom, crowding the aisles and making movement impossible. Employees don't know what inventory is available and can't find what they need even if they do know because it remains in stacks of boxes. Customers are unhappy that they can't buy the items they want when they want them, on their visit to the store. Employee morale is suffering because those who have in the past processed shipments resent the fact that others aren't contributing to this task.

- *What is causing the problem?* The answer is perhaps less "what" than "who": the manager herself. She may recognize that she needs a plan to allocate responsibility and, thus, asked you to investigate. But it may also be that an increase in inventory has put new pressures on processing, or that an increase in the number of customers has meant that the current staff, who usually had time to put things away, no longer has that time.

- *What is being done now to solve the problem, if anything?* Not much is being done. Some of the longer-term employees are making personal notes about boxed items of particular interest for their repeat customers, know where to find the item, and unbox it when needed. But that's not a sustainable process long term and only builds resentment among employees.

- *What's needed to solve the problem?* In conducting your investigation, you'll seek answers to such questions as: Where and when should shipments be processed? In the storeroom? On the loading dock? Every day? On shipment days only? At night? Who should be assigned to processing shipments? In a rotation? Should one person be relieved of other duties for this one?

- *What solutions are possible?* The solution, as indicated by the request from the manager, has to be a plan for deploying employees to process the shipments. Determining the extent and formality of the plan may depend on a look at the short- and long-term benefits of effective processing. Is it a good thing on its own? Does it contribute to better morale at the store? Can it improve customer service?

- *Which solution is best?* Within the limits of the store's budget, hours of opening, and total staff, the plan must allocate tasks fairly and reasonably to employees who will process shipments.
- *How and by whom should the solution be implemented?* This last question cycles back to an earlier one: who caused the problem? In this scenario, the manager recognized the problem and asked you to come up with a plan to solve it. She knows there's a problem; you have a good idea for solving it, and she'll be happy to hear it.

This sequence of questions and answers can lead naturally to a process for gathering information. At the same time, it can also embed an appropriate structure for arranging content in a final communication product. The questions start with defining the problem as well as need and move on to discussing the possibility of some kind of action to solve the problem. Answering the last question forces the choice of a solution, something specific that can meet the need.

To the degree that the problem is known to your audience, you can probably cut short the answers to the problem-based questions and focus on a solution. You argue that your management plan meets that need. But let's change the scenario. Imagine that you, as an employee, have become frustrated with the haphazard processing of sporting goods in the store. You are an orderly and responsible person, and so you've taken on the processing task yourself, squeezing it in while you fulfill your other responsibilities as a salesperson. You think that's not fair. You assume there must be a better way. In this scenario, as you communicate with the manager, you need to pay more attention to describing the problem in order to convince her that the problem and need exist. And you have to avoid seeming to blame her for overlooking the situation. You appeal, perhaps, to a higher value, such as your interest in contributing to a well-run store, which should please her. Your communication has to do more than explain and justify: it has to persuade her to act in the way that you think is in her best interest—and yours (Chapter 15 reproduces the student's brief implementation report).

IMAGINING OPPORTUNITIES

Thus far, this chapter has centered on defining a problem as a way to craft a solution. Doing so starts with answering the question: *Who knows?* Figure 5.1 sorts through four situations that pose that question, from the relatively easy (the problem is known to the person who needs to have it solved, and the solution is available) to the wildly imaginative. One way to investigate and define a problem is to categorize it, which helps you understand features it may share with other problems in the category. That understanding will help you craft options for solving it. You may also go through a sequence of questions common to problem situations, a heuristic, to discover more about the problem and, thus, how to match it with a solution. The sporting goods store employee faced a problem of means (how to process shipments) and applied questions to generate a solution (a management plan).

Turning Problems into Opportunities

The rest of this chapter turns toward the imaginative thinking inspired especially by the last situation described in Figure 5.1, which poses two unknowns. Design thinking can help in any problem solving situation, but it moves to front and center in addressing unknowns. It helps foresee problems unknown to others, questions that may be buried in answers taken for granted, that can become opportunities for innovation.

Creative Disruption

Broad changes in technology have opened opportunities to rethink and, in current terms, disrupt long-established and institutionalized sectors of the economy. Empathizing with potential hosts who might gain extra income by renting an unused space and with potential guests seeking an alternative to commercial hotels, Airbnb, for example, has disrupted the hotel industry. Lyft and Uber have disrupted the transportation industry—both taxis and rental cars. In doing so, both enterprises have contributed to—and built their platforms on—what has been called the "sharing economy": homeowners sharing vacant rooms or second homes; car owners, not professional drivers, sharing time in their cars. Even before the COVID-19 pandemic that forced many knowledge workers to work from home, the office was added to this list of disrupted economic sectors. Mobile ICT has led to changes in the concept of work and decoupled new forms of work and workers from the fixed office spaces of corporations. Observing and empathizing with this new demographic of creative professionals who can now choose where and when they work, entrepreneurs developed a new kind of office, coworking spaces, to fit their need.

These examples of disruptions in a creative economy suggest the big picture for seeing opportunities where others didn't. They depended on visionary founders who pulled on their own experiences (for example, spending time sleeping on friends' couches) to inspire them. Going big with ideas is something you can engage in, too. You may already be doing so. Carefully observing your surroundings—not taking things for granted—is one source of new ideas. So is collaboration, particularly in groups or on teams that bring together a diversity of interests, perspectives, disciplines, expertise, cultures, and languages. An individual scientist in a lab or in the field is still often the source of significant new knowledge, particularly in basic research. But in today's marketplace of ideas, collaboration is considered central. As one entrepreneur remarks, "Life is not about being in a shiny office building . . . It's about being in a zone where you're trying to figure things out the world doesn't know it needs yet" (quoted by Hu & Haag, 2020, p. A24).

Scenarios: Health Care

Consider another significant sector of the economy that offers opportunities for innovation worldwide: health care. The COVID-19 pandemic only increased interest in finding better ways to provide services, particularly through technology. Interdisciplinary teams of students have been taking on projects directed to this need. For example, one team at the University of Delaware developed, over several semesters, what they call a "smart boot" to help individuals who have restricted use of a broken leg know how much pressure they can place on that leg during their recovery. This orthopedic walking boot uses sensors, microprocessors, and rechargeable batteries to provide up-to-the-minute visual feedback that alerts patients if they are under- or overloading their leg. It is a formerly unknown solution to a known (to the suffering patient) problem (Kukich, 2016).

Here are a few other strategies for innovating. Focus, for example, on a demographic. One particular demographic of interest is the growing population of homebound elderly citizens who may need various types of medical assistance. The current approach often consists of visits from care providers. But such visits are at best only intermittent, and the alternative for more consistent, 24/7 care, such as live-in assistance or institutionalization, is even more expensive. In addition, institutionalization would be much harder on the patients who, by and large, would rather be at home. Caregivers, empathizing with their patients, want to do the best for them. In particular, they are looking for devices (like the smart boot) that enable patients to manage their own personal health safely and independently.

The growing demographic of the elderly presented the problem: there are not enough caregivers to serve them in the traditional way. Smart thinking developed the answer: innovative use of technology. In Finland, which produces a large number of digital health start-ups, entrepreneurs have developed various devices that patients wear to enable care workers to monitor them remotely. They also built robots. One holds a 2-week supply of multiple drugs that it distributes to patients at home at the appropriate time and records that distribution, otherwise the most common reason for a home visit. The robot chimes when it's time and dispenses the right combination (*The Economist*, 2020). To encourage healthy eating, researchers have developed a sensor-enabled dinner plate that can weigh and analyze chemically the food placed on it, obtain nutritional information from a US Department of Agriculture (USDA) database, interact with a caregiver online about health goals and recent intake, and flash green if the food is OK or red if not (Brennan, n.d.).

In these situations, researchers saw a need being met in one way that could be better addressed in another. It wasn't a problem—until something changed, such as the growth of a population of the elderly and emerging technologies—and then it became an opportunity to investigate.

Generating Innovative Solutions

You and your team can effectively adopt a similar attitude through design thinking. Start by empathizing with and understanding an actual or potential customer, client, or user. Continuously refine your definition of a problem or opportunity their needs might present. Frame and clarify potential solutions. Assess your resources, both within and external to your team.

In reviewing what you know and collecting information, you *create* multiple choices about the further conduct of your project and possibilities for its final outcome. You welcome and talk about all the many ways your views on the team might diverge. Along the way, you also *make* choices toward converging on the most promising ideas. In doing so, avoid the common traps in decision-making described in the box.

For more on negotiating team conflicts, see Chapter 4

COMMON TRAPS IN MAKING A DECISION

Consistency

Being consistent in a line of reasoning is important in a final product, but it has a downside in making choices when a commitment to consistency over-rules critical new information. The greater the commitment, the harder it may become to reverse course or abandon an approach, which may, actually, be the right action. Sometimes, you have to turn around.

Familiarity

What you know, or think you know, may hinder your ability to see the new. This trap, perhaps more common among seniors than freshmen in an academic environment, is thinking that the way we've always done something is the way to do it now.

Expert Halo

It's easy to assume that experts know what they are doing, and so it's safe to follow. That's often true, maybe more often than not. But especially in projects that require a broad interdisciplinary approach, expertise may have to be established rather than assumed. The role of expert may also change over the course of the work. The expert may also be a client with fixed ideas about what's wanted. Some members of your team think you should go along; others feel a different approach would be in the client's best interest. The situation will require negotiations.

Scarcity

In general, what's scarce (such as getting an audience's attention) is more valuable than what's abundant (too much information vying for the audience's attention). Launching out on a radical idea to gain attention may risk the team's reputation or financial status. The risk may be worth it. But when time, labor, and money are involved, the team needs to calculate the nature of the risk.

Acceptance

This trap is yielding to peer pressure, giving in to what the majority wants when it's not the right idea. You settle too soon on an answer because team members are bored or tired.

Groupthink

Unable to find healthy ways to disagree and air conflicting ideas, the team may settle on a conservative, mainstream idea that no one finds objectionable but that lacks innovation. Some members may comply with the group's decision out of fear or dependency rather than from conviction that the decision is right. Members of a group may tend to accept bad ideas—or, at least, ideas they don't really agree with—either because the ideas are presented by others with greater authority or because members of the group simply don't want to spend time on a contentious issue. Overemphasis on conformity leads to groupthink, which produces automatic responses, unexamined assumptions, and tired ideas.

Withholding of Effort

If members don't feel comfortable with each other or resist the authority of the team leader, they may withdraw from the discussion, content themselves with leaving things to others, and do less than they would if they had tackled the problem themselves.

Withholding of Information

Similarly, team members may refuse to share their ideas and information with the team. That refusal may

- Reflect a personal or cultural value in not wanting to confront others and speak up publicly
- Reflect their fear of making a mistake
- Reflect a need to save face
- Respond to their unwillingness to share credit for an idea they consider wholly theirs
- Derive from an unwillingness to disclose proprietary information.

If team members hold their cards too closely, the effort collapses.

Source: Adapted from Julavits (2019).

Avoid these traps as you collaborate to converge on an idea or two that seem to be the best solution in the situation.

GATHERING RESOURCES

The next two chapters provide strategies for collecting information to help generate solutions. Chapter 6 looks at empirical strategies, observation of things firsthand. When researchers write up firsthand accounts of such direct observations, that account is called a *primary* source of information.

All other sources of information are called *secondary sources*, a vast trove of knowledge. Chapter 7 briefly reviews strategies for accessing such sources, either online or in print. Secondary sources analyze, interpret, or explain information derived from a primary source or from other secondary sources. Research published in journals and books; in-house documents produced by companies, institutions, other organizations, or professional associations; newspaper and magazine articles; and other media, including social media: these are typical secondary sources.

You gain from these sources not only information, but also insights into the variety of media, structures, and styles in which you turn information into content to be delivered to your audience. This chapter has set the stage for managing a project through its focus on defining a problem or opportunity.

> In this chapter, you've learned ways to define problems and pursue the projects they set in motion. The problems may be ones that are assigned to you or are of general interest to you as a student. They may also be problems you'll encounter as a professional. Some of these problem situations may yield to simple solutions; others may be messy and complex. The strategies described in this chapter will serve you well whatever the situation, including those that offer new markets for creative products and services, even, perhaps, new opportunities to make the world a better place.

CHECKLIST: DEFINING THE PROBLEM OR OPPORTUNITY

1. Define the problem
   To launch a project to investigate a solution
   To see an opportunity for innovation and entrepreneurship

2. Ask the question "Who knows?" and consider four situations for problem solving
   Known problems with known solutions require routine practices
   The three other situations require increasing creativity in their solution
   Known problem—unknown solution
   Unknown problem—known solution
   Unknown problem—unknown solution

3. Categorize the problem you are investigating
   Problem of fact: What is?
   Problem of means: How to? How come?
   Problem of value: Should we?

4. Use a heuristic to gather information that also gives a preview of a draft document

5. Implement design thinking activities to generate opportunities for innovation

EXERCISES

1. Explore design thinking using the extensive resources available online from the Stanford d.school: https://dschool.stanford.edu/resources

2. Design thinking begins in empathetic observation of people interacting with their environment. To expand your ability to *observe* through all the senses, read the article in this link: https://bit.ly/2R4Jtz1

   Then, take on what a student calls an "empathizing and observing scavenger hunt" (Rebekah Smith, 2021). Spend 20 minutes walking around campus (or another location) and observing the scene. Think about ways the designs of objects, spaces, and processes impact people and try to empathize with the users' experiences. Take a picture of a design that caught your interest and bring it to class, so you can explain what you've observed and the opportunity you see for defining a problem. But, do more than just seeing with your eyes. Pay attention to surface textures, smells, and sounds as well. Use that experience to understand how you might craft communication strategies that appeal to different senses, beyond just sight.

3. You are probably learning to solve different kinds of problems as they are defined in different disciplines. Write a brief description of how one course or, more broadly, one discipline you are studying looks at problems. Use the categories of fact, means, and value to frame your description.

FOR COLLABORATION

In a small group, use the slide set in the link below, along with Zoom access, to run and participate in a 3-hour workshop. Part of the d.school starter kit (https://bit.ly/3e1Aojw), the workshop includes videos as well as guidelines for collaborative activities and discussions, including "over 20 concepts and tools that will help you apply human-centered design to your own work."

REFERENCES

Brennan, P.F. (n.d.). VizHOME: A context-based health information needs assessment strategy. Proposal.

Broadhead, G.J., & Wright, R.R. (1985–86). Problem/solution cases in technical writing. *Journal of Advanced Composition, 6*(2), 79–88.

Dusenberry, L., Hutter, L., & Robinson, J. (2015). Filter. Remix. Make: Cultivating adaptability through multimodality. *Journal of Technical Writing and Communication, 45*(3) 299–322.

The Economist (2020). Finland: Prescribing Tablets. January 11, 44.

Hu, W., & Haag, M. (2020). Why old industrial lofts draw shiny new tech firms. *New York Times*, January 23, A24.

Julavits, H. (2019). What I learned in avalanche school. *New York Times Magazine*, December 31. Retrieved from https://nyti.ms/3tDSEoq

Kukich, D. (2016). Smartboot: Engineering students add high-tech function to low-tech ortho-
    pedic boot. Retrieved from https://bit.ly/3n6mTCf
Penn Engineering's New Dean. (2015). *The Pennsylvania Gazette*, May–June, 26.
Smith, R. (2021). Personal communication.
Waber, B., Magnolfi, J., & Lindsay, G. (2014). Workspaces that move people. *Harvard Business
    Review*, October, 69–77. Retrieved from https://bit.ly/3sCQAfh
Zeeberg, A. (2020). Superthin and all powerful. *New York Times*, January 7, D1.

6 Generating and Evaluating Empirical Information

The student employee you met in Chapter 5 faced a problem in processing shipments in the sporting goods store where he worked. He asked a series of questions to help him find a solution. He observed current practices as well as the physical locations where they occurred and talked with others at the store, including his manager. His approach is a simple example of collecting data through what are called empirical methods:

- Experiencing things firsthand in environments of interest
- Conducting controlled experiments in the lab or in the field and building prototypes
- Gathering information through interviews and surveys.

This research approach is sometimes called *on-the-ground*, *field-initiated*, *experience-based*, or *inductive*. It is fundamental to projects in design thinking.

As you collect and analyze data, you come up with a statement, a guess, that aims to explain them. Composing that statement may reveal gaps in information and inconsistencies, as you learned in Chapter 5. The statement helps you determine what to do next: experiments to conduct, sites to visit, people to talk with, or further resources to examine to gain evidence for your explanation. In other classes and on the job, you've learned or will learn about the specialized empirical and experimental methods appropriate to research in your field. This chapter focuses on the communication dimensions of empirical investigations, both those aimed at exploring phenomena and those gathering evidence to validate a conclusion or hypothesis. It begins with a discussion of the ethical guidelines that protect the people whose interests and activities may be impacted by your research—and protect you—as you pursue your project.

FOLLOWING RESEARCH GUIDELINES

Design thinking begins in empathetic observation of others. That means you'll often be interacting with people as the subject of your research. When you do, you need to comply with appropriate guidelines for such research. The guidelines usually reflect state and federal mandates as well as those of your profession or organization. The guidelines address the need to protect the privacy and agency of your subjects and gain their explicit permission to be part of the study.

DOI: 10.4324/9781003093763-8

To comply with the guidelines, you submit a plan for your work. In academic settings, an institutional review board (IRB), which must approve the plan, specifies the plan's structure and content. It usually asks how you will select your subjects, what methods you will employ, and how your methods will impact them. The IRB may also require you to submit consent forms signed by your subjects. A range of templates for such forms is available online. You'll also need to establish who owns the final data or product from the research.

In general, the extent of the review reflects the degree of risk subjects face in the study. Tests of drug treatments or other procedures that affect subjects' health demand extensive reviews, usually with strict oversight built into the study itself. Less risky studies may receive an expedited review. Research in the classroom may be exempt from review. Even so, you should file a request with the IRB and always disclose the purpose and procedures of the study to your subjects to make sure they understand the conditions of their participation.

EXPERIENCING

Let's assume that, from personal experience, you know the time you spend out-doors makes you feel healthier, better able to deal with stress, and happier than when you've had to spend a lot of time indoors. Your teammates agree. You guess this health effect is generalizable over a larger population. How might you launch a project to find out?

The following discussion, which provides a brief look at some empirical research projects on the large and increasingly popular topic of nature and health, serves in part to frame this chapter's advice. It may also inspire your own thinking about ways to engage in a related project on campus or in your community. Some of these projects required extensive expertise, sophisticated instrumentation, and financial investment. But not all. Mostly, they required a recognition that something often taken for granted—being outdoors has health benefits—can be subject to investigation and proof. Such research can also lead to strategies for persuading people to spend more time outdoors and can uncover opportunities for new enterprises.

Researching Nature and Health

Here are some examples of health-in-nature research. (The following discussion is adapted from Morris, 2021.) The Japanese have a practice called "forest walking." To test its benefits, a researcher studied 12 healthy men who took two walks, each 2 hours long, on 1 day, in a forest park. In a follow-up test, the researcher found that the anticancer cells of the participants were significantly increased—a major benefit. In a study at Stanford University, a researcher asked two groups of students to walk for 45 minutes. One group walked through surrounding hills; the other walked on a busy street, but lined with trees. The researcher found that the hill walkers performed "dramatically better" on a series of cognitive tests afterward. "A 45-minute walk in nature can make a world of difference to mood, creativity, the ability to use your working memory." A pediatrician concerned about her patient's lack of access to the outdoors coordinated a trial study with a group that conducts nature outings. At yearly visits to her office, she asked children if they had access to the outdoors. If not, she referred them to the nature group for outings in nature

1 day a month. She found that each outdoor visit "decreased parents' stress and increased children's resilience." In a broad approach, a researcher surveyed 20,000 participants, who reported an outcome of "good health and well-being" when they spent 120 minutes or more a week in nature (less time didn't make a difference).

Engineers are using satellite temperature maps, pollution and census data, and GPS mapping to quantify the health benefits of green spaces. Environmentalists are experimenting with trees. In one project, researchers are planting full-grown trees in a schoolyard to learn what type of canopy is required to reduce air pollution and asthma. In another initiative, researchers in a design lab in California, along with other scientists, are studying the biophysics of vegetation in neighborhoods in Louisville, Kentucky, to test the potential for urban greening. One project aims to measure the impact on residents' health, including asthma, heart disease, and dementia, before and after the planting of 8000 trees in their neighborhood.

The design lab founder plans to launch a start-up based on his idea and subsequent findings. Another start-up is developing an app that will track outdoor time like apps that track steps taken. The company is also creating vegetation heat maps that combine data and infrared technology to show, on a scale from "nature rich" to "nature deficient," how healthy a neighborhood might be.

Engaging the Senses

Empirical research is often called *observing*. That shorthand term emphasizes seeing, which is clearly important, but limited. Research into such experiences as the health effects of being outdoors needs to tap into each of the generally recognized five senses: touch, sight, hearing, smell, and taste. In addition, being outdoors evokes another sense, the sense of being in *space*. Gardeners are implementing designs based on the smell of the vegetation, for example, to appeal to those who cannot see. Based on research, museums are incorporating appropriately themed music and items that can be touched into their exhibits to expand their audiences.

From their various disciplinary perspectives, the health-in-nature researchers walked around to experience outdoor environments themselves. They asked people questions about their experiences and designed studies to examine relationships. They also sought to improve the lives of their subjects through new technologies. As the developer of the outdoor-time app noted, "People are deciding whether or not this type of coffee bean or that type is better for you, when there is such an obvious health tool at your disposal. You literally just walk outside. People don't know" (Morris, 2021, p. A11). The researchers did know, or guess. They looked beneath the obvious to ask questions, observe more deeply, and conduct experiments. They saw what others overlooked.

Believing Is Seeing

In using observation as a tool to define problems and investigate new opportunities, keep these perhaps well-worn thoughts in mind. The statements focus on seeing but have larger implications:

Much of what you see depends on how you look.

(Lewis Carroll)

It's not what you look at that matters. It's what you see.

(Henry David Thoreau)

What you see and experience is often conditioned by what you have been trained or have grown up to see or feel. Your observation reflects your discipline, your profession, and your social and cultural background. Members of different groups and cultures may experience the world they live in differently. It's not just that they interpret the world differently; the evidence itself given to them—to you—as members of different groups is different. For example:

- *Customs*: Archaeologists see Native American artifacts found in the US as items for study, collection, and display in museums. The Native Americans themselves experience these artifacts as sacred connections to their ancestors that should remain where they are.
- *Observations*: As observers of nature, native populations often see things that non-native scientists miss in their brief visits—for example, peculiarities in the mating cycles of marine mammals in the Bering Sea.

Making the Familiar Strange

You can learn to experience things differently if you immerse yourself personally and intentionally in another culture. That's one important benefit of studying or traveling abroad. But you don't have to travel far to make yourself aware of how people shop differently, for example, in another part of town or another type of store, or how students in different groups take different perspectives on campus life.

In addition to seeing how strange or unaccustomed activities can serve familiar needs, like shopping, another perspective is to see how familiar activities can be made new, as with the health-care apps you read about in Chapter 5. You find new ways to achieve tasks that seemed to be settled and routine. You deliberately make the familiar strange and subject to innovation.

The latest technology, however, may not be the most appropriate in the situation at hand. In an empirical research project at the Technische Universität Darmstadt (Germany) funded by the European Research Council, researchers have examined the usefulness and productivity of indigenous technologies that circulated in various parts of the world between 1850 and 2000.

Some of these technologies persist today in what can be called the Global South, outside Europe and North America. They often demonstrate that the latest technologies are often not the most appropriate. Economic and political crises may make charcoal stoves and horse-drawn plows a better alternative than more modern equipment. Imported bicycles, motorbikes, and cars have to be adapted by local craftsmen for expanded usability and longer life. The researchers are finding that, contrary to much thinking in the US and Europe, technological change is not a "linear and evolutionary process" (Project Global Hot, 2020). More sophisticated technology is not necessarily the better application.

CONDUCTING A SURVEY

In addition to experiencing a situation on the ground, or sometimes in combination with such experiences, researchers often conduct surveys of individuals or groups they are interested in. As you read earlier, one of the health-in-nature researchers used a survey to ask respondents about their experiences.

You've probably responded to surveys that ask you to, for example, rate a professor and a course, comment on how you were treated as a consumer, or indicate what you are looking for in a particular product. A survey often precedes proposals for change. Instead of using sensors, as you read in Chapter 5, researchers interested in social networks in a workplace, for example, might circulate a questionnaire asking workers to identify the people they interact with daily, weekly, monthly, quarterly, and yearly. Specific questions might indicate types of interactions of interest: conversing informally, seeking advice, brainstorming to innovate on a problem, making decisions, or looking for other ways to improve their work practices (with a fill-in word box after "other").

Rapid Surveys

New technologies are providing ways to conduct surveys rapidly and, thus, to enable faster decision-making. For example, at a neighborhood forum, residents were surveyed about their preference in options for fixing an intersection considered to be a high-crash location based on Department of Transportation criteria. Residents used a keypad to vote on such specific options as reducing the speed limit from 35 mph to 25 mph, increasing the number of street lights, installing more crosswalks, adding more sidewalks, adding a traffic light, and making the intersection a four-way stop. Electronic voting yielded instant results for further discussion at the forum. The town officers and traffic engineers also used the results to aid their design.

Another new format is a short "pulse survey" to take participants' "pulse" on a targeted topic of interest. Such surveys include only a few targeted questions. A company may use them to survey employees on a topic of concern. The shortness, along with a requirement to answer before logging into a work computer, helps motivate a response. Companies may also use a pulse survey to encourage customers to respond quickly to questions about a product or service by clicking on an emoji or check mark. These online "Happy-or-not surveys," as one producer calls them, gather real-time data analytics for marketing.

Formal Surveys

To conduct more formal surveys as part of a research project, first secure approval from an IRB or other entity that oversees your work. Make sure you have a process in place to inform participants of the conditions of their participation. Then, follow these steps:

1. Develop an effective questionnaire, the tool of the survey.
2. Select an appropriate group to respond to the questionnaire.
3. Distribute the questionnaire through the best media.
4. Properly interpret what the respondents say.

A survey must meet two criteria: validity and reliability. A valid survey is one that provides the information you need. Validity is a measure of the questionnaire.

A reliable survey is predictive—that is, its results can be replicated among a different group of respondents. Reliability is a measure of the sample population. This section briefly discusses how to compose and deliver a questionnaire, choose a sample population, and interpret your results.

Composing the Questionnaire

To ensure that your survey is valid, design and word the questionnaire carefully to avoid language that indicates a bias or might be misleading. For example, candidates for office, on campus or in larger political arenas, may structure questionnaires in a way that elicits answers favorable to their cause. In an even more ethically suspect practice, they may use the survey form, especially online or in a phone call, to disguise a "push poll," a deliberate attempt to change voters' minds. Survey respondents can be highly suggestible. The options given sometimes bias the results. For example, the forms a professor created to elicit student responses to a course listed a range of answers where the lowest point for each question was "very good" and the highest was "outstanding."

In addition to bias, another problem that reduces accuracy in surveys is misleading or ambiguous phrasing. To be effective, keep questions as concrete as possible. Avoid giving too much room for interpretation, which will reduce the validity and comparability of responses. Instead of asking "Do you eat regularly," which leaves open the meaning of "regularly," ask specifics: "How many times a day do you eat? Do you eat in the morning? At noon?" In addition, respond to changing social conditions. A recent example included these options in response to a demographic question about gender: female, male, non-binary, prefer to self-describe (with a fill-in word box), or prefer not to answer. Avoid using jargon or bureaucratic language that will be meaningless or ambiguous to respondents.

COMPOSING AN EFFECTIVE QUESTIONNAIRE

Opening Explanation

In an email message linked to an attached survey, in a recorded voice mail greeting that begins a phone survey, or in a covering memo to a printed text:

1. Explain your purpose, the role you are asking the respondent to play toward achieving that purpose, benefits the respondent might receive, and limits on the use of the resulting data (for example, that each response will not be personally identifiable, and results will be aggregated).
2. Assure the respondent that participation is voluntary (if it is), and that the person's privacy will be maintained.
3. Detail the estimated amount of time expected for a respondent to complete the survey, the deadline for completion, and any requirements for responding (for example, must be over age 18, must have visited a provider at the health practice circulating the survey in the last 6 months). Provide a link, too, to further information about the topic or about the project.

Questions

- Minimize the number of questions, with only one or two on each screen, to ensure completion.
- Number each for ease of reference.
- Place the easiest or most interesting questions first to build motivation to respond.
- Include only one idea in each question.
- Make questions easy to answer: yes/no, multiple choice, fill in the blank, or a Likert scale (noting what number is high—for example, from 1 = strongly disagree to 5 = strongly agree).
- Perhaps include a final question that elicits an open response.
- Pretest the form with a small group of potential respondents to weed out ambiguities and unproductive questions.

Selecting the Sample Population

To ensure that your survey is reliable, carefully select the sample population. Make sure individuals are representative of the larger group whose behavior or attitudes you seek to identify, and that you haven't stacked the deck to achieve the results you want. In seeking opinions on restricting the use of a wilderness preserve, for example, you skew the outcome if you circulate a questionnaire only to those who hold hunting licenses or only to members of a nature conservancy. A survey on the drug-taking habits of the elderly is biased if only those in nursing homes are questioned. In addition, make sure the sample population is large enough to warrant the interpretations and trends you ascribe to it.

Distributing the Questionnaire

Multiple channels are available for distributing your survey. Choose the one (or more) that your participants would prefer. You can:

- Post it to a company website.
- Post it in a patient or customer portal.
- Send it through email, either within the body of a message or attached.
- Message it in your workplace collaboration software if it aims at that restricted demographic.
- Circulate it on paper, a still-common approach in some US medical offices.
- Send it online (especially as a pulse survey) in response to an online purchase.
- Create a social media poll (only if your survey is really short).

The options are expanding. Spend time considering which is the most likely to motivate a response. Your rate of response is a strong measure of the reliability of your results.

Interpreting Your Results

Survey research requires special training in the social sciences and statistics to select the best approach and interpret results accurately. If you are not familiar with statistics, for example, you might want to make sure that someone who has such skills is on your team. In addition, conducting surveys across cultures presents special problems. Americans are generally comfortable responding to questions, but you may find resistance or fear when you administer a survey elsewhere, particularly if you ask about such subjects as health practices and personal values that are not generally discussed in public. More than merely translating the words in a questionnaire designed to elicit health information, for example, you may need to work through a local agent who is trusted by your population and gathers information indirectly.

INTERVIEWING

Another method for gaining firsthand information is to talk with people in an interview. Interviews take many forms. Your conversations with other students or researchers constitute an informal interview. Through these conversations, you find out what others are doing about topics or problems that interest you, you test your own thinking, and you glean current, often suggestive, information that isn't otherwise available. In general, you aim to create new knowledge, not just to find out something that already exists in the literature. Tap into the grapevine in an organization and you may gain better information than what's provided in official statements. But, as with conducting a survey, ensure that the people you interview as part of a research project are well aware of the conditions of their participation.

GUIDELINES FOR CONDUCTING AN INTERVIEW

Before the Interview

1. Select your interviewee carefully. Don't waste your time—and theirs—interviewing someone who can't tell you what you want to know, even if you like them.
2. Do your homework. Know the basics about the topic of your interview and the person. You'll be resented and waste time if you inquire about facts you could easily have learned before. Get all that out of the way in advance.
3. Prepare questions.
4. Request the interview well in advance. In your request:

   - Provide options for conducting the interview, if appropriate (phone, video conference, or in person).
   - Suggest several times to suit the person's schedule.
   - Establish your own credentials (if the interviewee doesn't know them).
   - Explain why you have chosen this person.

- Suggest the topics you'd like to cover, maybe with a sample question or two.
- If you're planning to use your cell phone or other device to record the session, ask for permission to do so.

In the Interview

1. Arrive on time, in person or in the appropriate medium.
2. Take a few minutes to establish a cordial relationship and encourage trust. Let your nerves settle. Reiterate the reason for the interview.
3. Listen. Be flexible in adapting questions in the moment if a new line of questions looks better than what you had planned. Use what the person says as the springboard for your next remark.
4. Record the session, if you've gotten permission, or take notes, sparingly. Take down statistics or other precise data, but don't try to capture every word.
5. Verify key or controversial statements before you end the interview.
6. If you are conducting a series of interviews, ask questions rigorously in the same order and avoid offhand remarks that might prejudice the comparability of results.

After the Interview

1. Compile notes or transcribe the recording while the interview is fresh.
2. Thank the interviewee in writing (if that's how you arranged it) or informally.
3. When you incorporate the interview in a document or other communication, refer to it by a citation that includes the name and title of the person interviewed, the place or medium, and the date. The date is needed to establish a context; it will protect you and the interviewee should their opinions subsequently change.

For more about participating in a job interview, see Chapter 12

The box on "Guidelines for Conducting an Interview" provides guidelines for interviewing a participant in a pre-arranged meeting conducted either in person or via phone or video conference. Read up on the person and topic first. Start with a list of questions, avoiding those that lead to yes/no answers or stock descriptions. Prefer ones that investigate the interviewee's response to the situation of interest by eliciting stories, specific instances, or perspectives. At best, your questions will help the interviewee think in a new way about the situation. Be prepared to abandon your list if a more productive line of questions emerges. Listen actively for such a shift in topic and pivot quickly in response. Watch for cues to potential questions or insights in the interviewee's gestures, dress, and physical environment, either where you are together, in person, or in the background of a video meeting.

A form of group interview is a *focus group*, a technique common in marketing research. In a focus group, six to ten customers or clients meet with a facilitator to discuss some product or policy currently in development. Groups are selected on the basis of similar characteristics. To confirm the accuracy of the findings, two groups per characteristic are usually required. The discussions last 1 hour or so. As participants listen to each other, they are reminded of thoughts or memories that can enrich the data.

PARTICIPATING IN COMMUNITY-BASED RESEARCH

In each of the methods described so far in the chapter—experiencing, conducting a survey, and interviewing—you as the researcher direct the process. Grounded in design thinking and an instinct for community, some research is taking a different turn. User experience (UX) researchers, for example, have developed a method they describe as "listening with your eyes and heart" as they evaluate the appropriateness of documents to their audience. To arrive at insights about how users experience documents on site, in the place of use, researchers observe (with their eyes), they talk with users (listen), and they interpret what they see and hear through empathy (with the heart).

For more about UX, see Chapter 11

In a further extension of an empathetic methodology, sometimes called "participatory visual research" or "community-based research," you let the research subjects themselves take center stage. You intervene less. This shift in research perspective is suggested by a shift in language. No longer "subjects," these individuals become "hosts," and the researcher is a community member, too. The hosts create visual representations of their interests and perspective and tell their own stories, in person and in audio files. This method for creating knowledge is driven and directed by the community whose interests are at stake. They are agents with you in the project. You learn from them while they learn from you, creating new shared knowledge through a process of dialog. You adjust your ideas incrementally in response to unexpected happenings during your fieldwork. At a simple level, you walk around, observe, and talk *together*. Incorporated in service-learning projects, participatory research enriches academic–community relationships.

The term *cellphilm*, which combines *cell phone* and *film*, has come to designate a range of research practices that are community-based, interdisciplinary, and multimodal (Macentee et al., 2016). The major tool is the cell phone. Participants photograph, speak about, and video their world, from their perspective. To determine what it feels like to work in a particular office environment, for example, a researcher could ask a sample of workers a series of questions or photograph them at work. Instead, in a recent investigation, a researcher asked workers to take pictures, from their perspective, of key elements in their workspace. The photographs became the source of questions, primarily about why the workers chose the items they did and their angle of vision on them.

This approach helps democratize the research process and elicit questions that reveal insights into how individuals imagine potential future change in their own work environment, neighborhood, or other community. Gaining such insights is fundamental to understanding the culture and behaviors in a community and advocating for beneficial change. In areas, for example, that are prone to such environmental risks as sea-level rise, wildfires, or pollution, gathering resident stories about their experiences on-site can provide important data about strategies for mitigating those risks. In addition, the information in itself can be persuasive. When the residents tell their own stories, their audio files or videos enhance the authenticity of what they say. They also do so in a medium that attracts attention. The stories provide information and motivation for neighbors and others to behave appropriately.

For more about risk mitigation communication, see Chapter 8

EVALUATING EMPIRICAL DATA

The goal of your research is data—that is, facts and statistics that you interpret to help solve a problem or make sense of a situation. The information serves to plug a hole in your understanding of a concept, phenomenon, or event and to make things more certain for yourself and for the people your project aims to serve. But not all the information you gather will be useful or right.

What Counts as Good Data?

What are *good* data? The answer is complex. In research that is highly quantitative and involves technology-aided observation, once you've made sure the technology is working, you can probably trust the data. But the data also need to be understood in context. For example, an iPhone app "designed to detect car-driving jolts and automatically report potholes in Boston was abandoned after the maps that were generated favored young, affluent areas where more people owned iPhones and had heard about the app" (Shaywitz, 2021, p. A15).

Evaluating data requires you to recognize your own biases, keep an open mind, and corroborate your findings with those of other researchers. As a historian of science argues:

> Everything we know is known by us; we can't eliminate ourselves from the picture. Defining methods, choosing which ones to use, deciding how to use them, understanding what they produce: each of these acts is fundamentally interpretive.
> (Riskin, 2020, p. 50)

The data you collect reflect your own experience and point of view, which may lead you to overgeneralize a concept, such as the health benefits of being outdoors. It also depends on your not being too invested in an initial hypothesis or explanation for an occurrence or phenomenon. Given the abundance of data available to you, you may be tempted to select only those that support your idea. You ignore the rest. This tendency is prevalent enough to have a name: *confirmation bias*. If you're working in a particularly uncertain problem area, such as sustainability or disease mitigation, you need to maintain a learner's perspective. Keep in mind competing hypotheses, alternative views, and unusual correlations. Be open to surprise and to changing your mind. Such multiple claims are a well-recognized underlying condition for technical and scientific investigations. The scientific method, with its emphasis on the reproducibility of results, aims to address this uncertainty and correct for confirmation bias.

Community-Based Data

Your position as a researcher is even more significant in qualitative research than in the more statistics-driven approach of data sciences. This is especially true in community-based projects where you are trying to balance people's different needs along with the needs of the physical environment. There's more room for interpretation. As with other empirical research, the process entails the gathering of data about the situation of interest and reviewing it to find repeated themes or patterns. In so-called "grounded theory" investigations, the themes are coded with key words and phrases, grouped into a hierarchy of concepts, and ultimately stated in a generalization or conclusion that pulls the evidence together, sometimes as a hypothesis for further testing.

The process is often conducted collaboratively to correct for individual researcher biases. Several people together

- Come up with key words for themes
- Discuss and prioritize the applicability of these key words
- Decide on the most productive terms
- Appropriately divide the evidence into those categories
- Compose the generalization that best explains the evidence.

The goal of this process is to achieve *inter-rater reliability*, an agreement among experts on what the evidence means.

Enough Data

All research raises questions concerning saturation—that is, what criteria to use to know how much data you need and, thus, when you have enough data and can stop. As you experience and observe things, look for inconsistencies and holes, for whatever doesn't seem right. Keep searching until the inconsistencies are resolved or well explained as inconsistencies. And plug the holes. At least for now.

In this chapter, you've learned some strategies for gathering empirical information by gaining experience in situations of interest, observing people and places firsthand, conducting surveys and interviews, and participating in user-experience and community-based research. You've also learned to respect the agency and privacy of any people you engage with as you conduct your research. In addition, you need to be aware of your position as a researcher in the situation you are studying and the limits on your interpretation of the evidence you collect.

CHECKLIST: GENERATING AND EVALUATING EMPIRICAL INFORMATION

1. Follow appropriate guidelines for research with human subjects
2. Extend your method of *observing* to incorporate the five senses and space

Don't rely only on what you see ("seeing is believing")

Recognize that different people see things differently ("believing is seeing")

Investigate other senses in gaining information: touch, hearing, smell, and taste

Gain firsthand experience in a place of interest to your research topic

Keep aware of the possibilities, as well as the limits, of technology

3. Conduct a survey

Use short "pulse surveys" as well as formal ones, as needed

To help ensure validity in your results, compose an effective questionnaire

To ensure reliability, avoid bias in the sample and include a large enough sample

Choose appropriately from the abundant channels for distributing the questionnaire

Pay attention to statistical methods and cultural issues in circulating and interpreting your results

4. Conduct an interview

Select the interviewee strategically

Know appropriate background before the interview

Prepare questions, but be prepared to change tack if warranted

Verify key or controversial statements before closing

5. Participate in community-based research

Engage with users as they experience a prototype on site, in use (UX)

Let research subjects tell their own stories, using cell phones and audio files

Walk around a place of interest with subjects, creating knowledge together

Subjects become agents with you in advocating for beneficial change in their community

6. Critically evaluate your data and conclusions

Recognize your own biases and limits on your methods

Keep an open mind

Corroborate your findings with those of other researchers

Maintain a learner's perspective

EXERCISES

1. Write a five-question questionnaire concerning a policy of your college or university. You might focus on, for example, holidays, class schedules, hours of access to campus buildings such as the library or laboratory spaces, or parking privileges. Make sure you know the relevant policies, including policies about distributing questionnaires to other students. If that policy allows, circulate your questionnaire to ten students. Create a table to display your results (see Chapter 9) and explain your results in a brief message addressed to the authority responsible for the policy, assuming that the administrator has requested your study.

2. You may try to ignore the surveys that arrive in your inbox after you've purchased a product or a service online. But next time you see a legitimate one, click on it. Note the kinds of questions asked and how easy (or not) it is to respond. What incentives does the company give you to take the survey? A measure of time (it's short)? Some reward? For more about pulse surveys, read about them in this link: www.happy-or-not.com/en/

3. Conduct an *informational* interview with an expert on a topic you are researching. Refer to the guidelines in the box on "Guidelines for Conducting an Interview" and the surrounding discussion. Develop a set of questions that help you both select the right person to interview and obtain meaningful results. Write up your notes after the interview to make sure you have recorded information properly while your memory is fresh. Check back with the interviewee if you find gaps or ambiguities.

FOR COLLABORATION

This chapter opened with a discussion of empirical research on the broad topic of nature and health. Let that discussion inspire your own thinking about forming a collaborative project to investigate some aspect of the relationship between human health and nature. Design the project to require empirical research. If possible, compose an interdisciplinary team of students. Your class might provide a good population for assembling several such teams to pursue this exercise as a class project.

Each team can focus on a specific problem or opportunity through the strategies you learned in Chapter 5. In your team, determine appropriate empirical methods for generating information. One method will probably be firsthand experience in a natural setting of interest. Consider other methods detailed in the chapter as well. Write a brief description of a specific problem or opportunity you are investigating and the empirical methods you would implement to conduct research, especially methods that might reflect your team's different disciplinary perspectives. You don't need to actually implement the methods, but provide evidence about why you think these empirical methods would be productive. Discuss your description in class and note differences among the approaches suggested by each team.

REFERENCES

Macentee, K., Burkholder, C., & Schwab-Cartas, J. (2016). What's a cellphilm? An introduction. In K. Macentee, C. Burkholder, & J. Schwab-Cartas (Eds.), *What's a cellphilm?* (pp. 1–15). Rotterdam, The Netherlands: Sense.

Morris, B. (2021). For better health, just head outdoors. *Wall Street Journal.* February 16, A11.

Riskin, J. (2020). Just use your thinking pump! Review of *The scientific method: An evolution of thinking from Darwin to Dewey. The New York Review*, July 2, 48–50.

Shaywitz, D. (2021). Broadly informed, easily misled. Review of *The Data Detective. Wall Street Journal.* January 29, A15.

Project Global-Hot. (2020). A global history of technology 1850–2000. Technische Universität Darmstadt. Retrieved from https://bit.ly/3ngRzAR

7 Incorporating Sources
From Research to Communicating

This chapter continues the discussion of research methods begun in Chapter 6, which focused on generating information through empirical methods. This chapter looks at the second broad category of research strategies, often called *secondary*. In secondary research, you read what others have learned and argued about the subject of your investigation. The term "read" is shorthand for discussing how you interact with a wide range of sources in the multimodal network for knowledge creation in the 21st century.

Primary sources are the data recorded firsthand—for example, transcripts of interviews, tabulations of survey results, scientific data collected in open-access repositories, data collected automatically by instrumentation, and data gathered by algorithms on various social media platforms. Secondary sources present data that have been interpreted in an explanation or a claim, a hypothesis proven or to be tested. Secondary sources, in turn, can be divided into two large categories: those *internal* to where you work and study and *external* sources. External sources include ones available through social media as well as more traditional published sources, often called "the literature." As you turn from researcher to communicator, use the literature not only as sources of information but also as models of communication. Analyze in particular how technical and scientific articles deliver information to you, and then imitate and adapt effective structures and styles in your own composing strategies. The last sections of this chapter provide advice on using these models to write a technical article and a review of literature. The review can be one segment of an article or, sometimes, a document in itself.

INTERNAL SOURCES OF INFORMATION

When you're on the job, as an intern or a regular employee, company documents will probably be your main sources of information. These include:

- Annual and quarterly reports
- Financial statements
- Catalogs or other lists of products or services
- Project reports
- Correspondence (including email and other messaging)
- Policy handbooks
- Manuals.

DOI: 10.4324/9781003093763-9

Workplace collaboration software, either a proprietary application developed by the company you work for or one available commercially, such as Google Workspace, Slack, or Microsoft Teams, provides a platform for producing, archiving, and retrieving all such documents. The program can store these documents in their final or draft form. It can offer a content-management function that accommodates topic-oriented chunks of text or visuals that you assemble in new documents as needed to meet new situations. Such applications support the collaborative authorship common in organizations. Who wrote which segments of an assembled document is less significant than the responsibility an editor or final reviewer takes for making sure the product meets its purpose and its audience's purpose. Authorship is often corporate, not individual, and guidelines for reusing any information within these documents are usually set by the company.

For more about collaborative writing, see Chapters 4 and 11

EXTERNAL SOURCES OF INFORMATION

Information about science, technology, and related subjects of interest is vast and increasing, as you know. It's also easy to access using such search engines as Google and Bing. For more specialized searches, you may rely on databases that your university or college subscribes to. Companies, including publishers, want you to find their information. Eager to market their products, they maintain teams or hire consultants to enhance their ability to be featured prominently in the search engines likely to be used by current or potential customers. The process is called search engine optimization (SEO).

At your end of the process, sort through the options in secondary sources available to you. To do so, think about three large spheres in which information is communicated across borders of disciplines, genres, communities, and forums (adapted from Applen, 2020). One sphere, the source of the most reliable scientific and technical information, includes the journals, other publications, and websites of professional organizations and specialized publishers. The second is the public sphere of information produced by organizations and generated by users on social media. The next two sections of the chapter take a closer look at these two categories.

The third sphere centers on you, as an individual. Personal taste and preferences determine what you read and what connections you make to other individuals. Your choices are increasingly reinforced by the algorithms of the platforms on which you do much of your reading. This makes it efficient for you to get to favored sources. But the process can lead to confirmation bias, an attitude that hinders both getting at what is true and generating innovative ideas. For scientific and technical professionals, search algorithms have widened the range of journal articles available over a lengthening time frame. But a sociologist discovered that researchers, "beholden to search algorithms' tendency to generate self-reinforcing feedback loops, are now paying more attention to fewer papers, and in general to the more recent and popular ones—actually strengthening rather than bucking prevailing trends" (Beckerman, 2018, p. 14).

In a similar self-reinforcing loop, you may decide that what is true is simply what others you trust in your community believe is true, without an independent investigation. The others may be wrong. Confirmation bias can present a powerful downside of community-reinforced thinking.

The problem you face in using any of these sources is knowing which sources are reliable. You need to sort valid information from *misinformation*—that is, errors in information or reasoning, often unintended—and *disinformation*—deliberate lies, often aligned with a strongly held opinion or belief. The following discussion briefly introduces some ways to evaluate sources. In addition, refer to such resources online as these:

- Newslit.org
- FactCheck.org

Technical and Scientific Literature

You'll find the most reliable technical and scientific information in the journals and on the websites of such professional societies as the American Institute of Architects, the Royal Geographical Society, or the Institute of Electrical and Electronic Engineers. The British Medical Association produces the *British Medical Journal* (*BMJ*) and *The Lancet*. Scientific publishers also produce such well-respected journals as *Nature*. The journals vary in prestige and coverage. A major measure of a journal's prominence and readership is its *impact factor*—that is, its relative impact on research in the field. That factor is calculated by such indexing organizations as the *Science Citation Index* (SCI), which traces, in effect, the offspring of an article by tracing all the articles in which a known article is cited. The higher the number of citations, the greater the impact is assumed to be. The SCI listings also indicate the cluster of research around a particular topic—useful information for researchers pursuing that topic. The journals mainly publish original research, although those addressed to practitioners, such as engineers and architects, also often include tutorials that show ideas in practice.

In the journals, researchers write for each other and have the expertise to evaluate, for example, the validity of highly specialized experimental techniques, the accuracy of results achieved, and the significance of those results in the larger picture. Authors debate hypotheses that are currently circulating and update each other. Both the authors and the readers of journals are researchers in the specialty of interest. Some journals, especially those addressing a small, niche interest, are available only online. Governments also issue technical reports, policy statements, regulations, maps and charts, census data, and records of such governing bodies as the US Congress. Increasingly, too, government agencies are maintaining large, open-access data sets to help scientists advance understanding of such basic mechanisms as the structure of the human genome.

A process called peer review has traditionally established the reliability of scientific and technical journals. Experts in the subject of an article submitted for publication read it for accuracy and significance before the editor accepts it. On the basis of the review, the article may be rejected or may have to be revised and resubmitted in a process that can take considerable time. Increasingly, especially during the COVID-19 pandemic that began in 2020, editors of health-related journals in particular felt pressure to disseminate findings faster to aid scientists and policymakers at the front lines of addressing the disease.

To address this need, some journals posted submissions online for access before review in an effort to enhance collaboration among the scientific community to "rapidly confirm and build on one another's findings rather than unnecessarily duplicating experiments" (Tingley, 2020, p. 18). Because such corroboration is the essence of the scientific method—making claims and amending them—that approach makes sense. But it challenges the role of journals as both gatekeepers of research integrity and sources of new ideas. Pre-review publication requires an understanding that claims are only valid in a context and at a particular time. Conclusions and recommendations need to be updated when new information is inconsistent with them, in an iterative process.

Public Forums

Beyond the sphere of technical and scientific literature, the second, wider sphere of information dissemination incorporates a broad public. Such information originates, for example, in organizations that produce technically oriented text and visuals, often at regular intervals, in a recognizable format, such as a magazine or newsletter, delivered in print or online. Public-oriented information is also increasingly found on social media, often generated by users, either in groups or individually. Information from organizational sources and user-generated content are discussed in the following sections.

Organizational Sources

Many public-facing sources related to scientific and technical information have a track record of providing valid information. These include, for example:

- The bulletins of agricultural and engineering experiment stations and extension services
- Business and financial publications (such as *Latin Finance*, *Business Latin America*, and *The Economist*)
- Research-oriented university publications aimed at a general audience
- Long-standing news sources, such as *The Wall Street Journal* and *The New York Times*.

In such publications, researchers report on their own work, or staff writers and journalists create original articles based on interviews and careful reading of the scientific literature. In reading these sources, which adhere to standards of professionalism and trust, be sure, for example, you're reading straight reporting rather than an op/ed (opinion or editorial) piece, which is usually flagged as such. Professional organizations and publishers also maintain a presence on social media. They offer public-facing expertise, in blogs, podcasts, videos, audio files, and other formats, as they welcome comments from their audience.

If you're not familiar with an organization producing the information you want to reuse in your own research, check it out. Enter its name, or the headline of the article you suspect is misleading, into a search engine to determine who sponsors the organization:

- Is it state-run?
- Is it aligned with a political or religious cause?
- Does the source provide original reporting or commentary on other reporting?
- Are there any disclaimers about the organization's purpose or origin on the website (often in small print)?
- Can you identify spelling or grammatical errors that might indicate a lack of professionalism in the source?

If something looks unfair or biased to you, try to corroborate that fact or interpretation by researching coverage in other reliable sources (adapted from News Literacy Project, 2021).

User-Generated Content

User groups are an important and ever-growing way to exchange information online. Social media are the town square or the town common where everyone can gather and have a voice about any topic. Birders can meet other birders, share their enthusiasm, and advance knowledge of birds all over the world. New parents share advice about parenting. Some COVID-19 survivors organized a group to share their experiences with each other. The posts also alert medical practitioners to an important source of data for future treatments. User-generated content on StackOverflow, for example, some in the form of stories, helps web and software developers face and solve problems (Lam & Biggerstaff, 2019). Advocacy groups urge collaborative action on such topics as the environment, social justice reform, or health and safety matters at work. Neighborhood groups exchange information about events of local interest, such local service providers as plumbers and electricians, and recreational opportunities and interest groups. Groups emerge out of a felt need at a particular time. The space for discussion expands indefinitely. In addition to their primary function of sharing information among users, these conversations also serve as a new form of fieldwork for researchers interested in what people do and say on social media.

The ability to capture and aggregate individual queries about diseases on social media platforms, for example, can advance global understanding of important trends and help mitigate serious problems. In a process now called *infodemiology*, Google Flu Trends has analyzed such queries to aid healthcare professionals in developing effective communication strategies, especially when there is a gap between what experts and what ordinary people think is causing the problem (Gilbert, 2020). Google Flu Trends was used by researchers to track the spread of Ebola in West Africa in 2014, the Zika virus in Brazil in 2016, West Nile fever over several years in Italy, and COVID-19 in 2020. During the 2020 pandemic, researchers found that searches for COVID-19 symptoms could track the spread of the disease several days in advance of official statistics, a technique referred to as *nowcasting*. Among the most predictive was the search for a loss of smell, something not originally considered. It became a key predictor built into a series of questions about symptoms asked by medical professionals to screen patients before providing services.

These opportunities to gain and circulate knowledge represent a very real upside of social media in research. The downside is that the space for misinformation and disinformation expands in equal measure. A spokesperson for the World Health Organization characterized this condition during the COVID-19 pandemic as an *infodemic*: a pandemic of false information. The lesson for you is to be careful when you reuse user-generated content in your own communication. Look carefully at visual evidence that may have been doctored or otherwise distorted to prove a point. Read comments on a post as evidence to either validate or contradict what the original post said. Trace the origins of a post to determine if it's actually an original or itself a repost. If the post cites sources, are they known to be reliable? As the News Literacy Project, a reliable source that advocates for "a future founded on facts," recommends:

> Learn digital verification skills. Use widely available free resources online to teach yourself how to do reverse image searches, use geolocation tools (like Google Street View), search for archived webpages and use critical observation skills to do your own fact-checking.
>
> (News Literacy Project, 2021)

INTEGRATING EXTERNAL SOURCES INTO YOUR DOCUMENT

As you do your work every day, you probably move back and forth between reading sources and writing your own documents on the same digital platform (Leijten et al., 2014). Take care to clearly distinguish someone else's content from your own. Identify external content in short modules of text or a visual keyed to a topic heading. A module may include one fact or a cluster of facts, your paraphrase or summary of one source, or a quotation (in quotation marks). Use abbreviations consistently to identify the source of each item at the end of the item itself.

Acknowledging the source recognizes who owns the words, images, numbers, facts, ideas, or points of view you select to include in your own document. One of the best tests for reliability in the information you read is knowing its origin. That identification includes, for example, the name of the author, the name of the journal, organization, or government agency that published it, and the date, as timeliness is a key value. Your own reliability is enhanced by citing reliable sources.

Quoting

Early in your composing, you may be tempted to quote everything. Resist that temptation. Gain confidence in using your own form of expression. Save direct quotation for rare occasions when it is especially economical, memorable, and authoritative. Use quotations to

- Call attention to a particularly significant interpretation
- Summarize with special finality a line of reasoning
- Present the accurate text of a law or regulation
- Open the discussion on a point of familiarity with the audience when the source is well known
- Capture a particularly vivid expression of an idea.

Visuals you want to reproduce should bear up under reproduction. A color original, for example, may not be as informative in a black and white reproduction. Change figure numbers and captions in the original to fit the sequence in your own report.

Paraphrasing

When you restate information from a source in your own words, you paraphrase. Use paraphrasing to condense or summarize a wordy or roundabout original and highlight an author's main point:

- *Original*: It is significant to note that the estuary has exhibited a dramatic recovery in water quality, especially in oxygen levels, compared to the conditions that prevailed 20 years ago. This is a result of wastewater treatment facility construction. It also results from the reduction of industrial discharges that has followed the enactment of such legislation as the federal Clean Water Act.
- *Paraphrase* (with source named as part of the sentence): To summarize, Guerreo (2021) testified that water quality in the estuary, especially oxygen levels, has improved greatly in the last 20 years because of enhanced wastewater treatment and federally mandated reductions in industrial discharges.

Acknowledging Proprietary Information

Acknowledging sources fulfills an ethical obligation to the people you write for and the people whose knowledge and language you write about. It also serves to establish your own authority as you cite sources whose credentials are well established. Beyond direct quotations, such proprietary information as the following is usually documented:

- Opinions and predictions
- Statistics derived by the original author
- Visuals in the original
- Author's theories
- Case studies
- Unique research procedures.

Documentation Guidelines

The reader for your writing (a professor, an editor, a supervisor) sets the rules for what and how to document. When you write in a class where you are learning to use external sources, expect to be closely watched. When in doubt, over-document. Professional journals require authors to let their readers know whose work informed the article at hand. As you'll see later in this chapter, a literature review segment confirms the thoroughness and validity of the author's research. It also points readers to original sources from which they can pursue another line of inquiry. Lengthy lists of references and footnotes are common.

Newspaper and magazine writers routinely give less than a full accounting of all their sources. Their readers trust that the situation is being fairly and ethically represented. They avoid the publication if it proves untrustworthy. They do not expect the same hefty apparatus of documentation as in a term paper. Outside the United States, with its intensely legalistic and individualistic view of property rights, standards for citing sources are sometimes more relaxed, except in the sphere of international scientific and technical publications described earlier.

Documentation Conventions

Every documentation system consists of two components:

- A brief citation next to referenced material in the text itself
- A full citation or "address" for each source in a list of references at the end of the text

Different organizations, government agencies, professional societies, and journals publish their own style guides that detail the mechanics of citation. These are available, often in updated form, online. Updates are frequently needed to accommodate the many hybrid genres emerging in both traditional publishing and social media. In doing your research, make sure you record all the publication information for a source, even if some information (for example, the author's full first name or the volume and issue number of a journal) proves not to be appropriate in a particular style. It's easier to eliminate an element than to recover a missing one.

Read the documentation rules that apply to your writing and then imitate examples of citations that conform to those rules. Citing sources is that simple. But it demands close attention to details. Two guides commonly used by technical students and professionals are those of the American Psychological Association (APA) and the Council of Biology Editors (CBE). Other popular guides are published by the University of Chicago Press and the US Government Printing Office (GPO). These are available on the organizations' websites.

ADAPTING A MODEL: THE TECHNICAL ARTICLE

Reading sources from the literature helps you learn the technical information you need to advance your project as you add value to others' content through your own research. Read these sources, too, as potential models for presenting your information. Imitating and adapting are excellent strategies for composing your own document. The following discussion focuses mainly on the structure and style of a technical article, but these strategies are also useful in writing other communication products (Leijten et al., 2014; Dusenberry et al., 2015).

Structure

Articles in professional journals offer a thorough, detailed account of an empirical or qualitative research project. They emphasize new and significant findings that make a substantial contribution to the discipline represented by the publication. Many articles are collaborative endeavors, and most are written by the researchers themselves. In their articles, as one seasoned author notes, researchers trade information for reputation. They lay claim to their discoveries in a field where many researchers are pursuing similar problems and argue for the truth of their assertions or hypotheses.

In writing an article, you show how your work supports, refutes, or extends established theory or confirms a new experimental approach. A *review of literature*—that is, your summary of the sources you gathered that are pertinent to your project—often appears in the introduction. It can also occupy an entire document. Writing such a review engages you in thoroughly understanding and reusing ideas from sources in your new context. For that reason, this chapter ends with advice for composing such a review.

The form of the scientific or technical article differs somewhat from journal to journal, but most follow the IMRAD approach, also common in reports. A report, however, often addresses managers, clients, or sponsors, while an article highlights what's news to colleagues. For this reason, the emphasis in each segment may differ. For example, a report may center on the description of a procedure, but, because such information may be simply common knowledge to an audience of colleagues, the article would focus on how the results contribute to the profession's understanding of key phenomena. If you wrote a report on your research that included several case studies, you might combine them to make one example in the article or highlight just one significant or interesting case. In general, however, the IMRAD structure works: introduction, materials and methods, results and discussion.

For more about IMRAD reports, see Chapter 16

Abstract

Published articles generally include an abstract, a summary of the major findings, usually about 250 or so words. Sometimes, the abstract appears separately as well, perhaps online to alert audiences in advance of final publication. Here is the abstract of an article titled "Story Mapping and Sea Level Rise: Listening to Global Risks at Street Level," by Sonia H. Stephens and Daniel P. Richards. You'll read part of their introduction shortly.

For more about how they planned their research, see Chapter 4

While interactive maps are important tools for risk communication, most maps omit the lived experiences and personal stories of the community members who are most at risk. We describe a project to develop an interactive tool that juxtaposes coastal residents' video-recorded stories about sea level rise and coastal flooding with an interactive map that shows future sea level rise projections. We outline project development including digital platform selection, project design, participant recruitment, and narrative framing, and tie our design decisions to rhetorical and ethical considerations of interest for others developing interactive tools with community participants.

(Stephens & Richards, 2020, p. 5)

Introduction

In the introduction, you create a "research space." You do so in three "moves":

Stasis—define research territory
Disruption—interrupt stasis so as to create a niche within territory
Resolution—occupy or defend that niche

(Harmon and Gross, 1996, p. 70)

Begin by setting the context—the static picture. Note prior research (that's the *review of literature*) to establish the scope and importance of the problem. Then, cite what disrupts that static condition—a gap, an inconsistency, a piece that doesn't fit the puzzle—and explain why that disruption is undesirable. Finally, note how or perhaps why your research resolves the problem—and lead the reader into the article or report to find out. Figure 7.1 shows these moves in action in the introduction of the story mapping article.

The authors continue by discussing the limitations of on-the-ground studies and introducing the "excitement and optimism" around advances in interactive

Stasis
Those tasked with communicating environmental risks have long grappled with how to most effectively engage public audiences on risks that have global, national, or local consequences. The traditional information deficit model of science communication, which posits that public audiences do not understand science data simply because they do not have enough scientific training, has been recognized as insufficient for communicating scientific subjects, including environmental risk (Bucchi, 2008).

Review of literature
Moreover, where risk has traditionally been defined by experts as the probability of a hazard's occurrence times the size of its impact (Okrent, 1980), contemporary research shows that public perception of risk is multidimensional (Slovic, 2010) and includes a risk's impact, an individual's confidence in scientific understanding of the risk, and perceived dread (Fischhoff, 2009). Peter Sandman (1993) captures this complexity succinctly in his definition of risk: risk=hazard +outrage. In light of this movement towards public inclusion in risk assessment and rhetorical framing of risk perceptions, the challenges of risk communication have become more clear: the data need not only merely be accessible and available but need to matter in a way that relates to people and their mental schemas (Lakoff, 2010), their environmental frames (Nisbet, 2009), their worldviews (Akerlof et al, 2016), and their sense of place (Scannel & Gifford, 2013)—in essence, their lived experiences.

Disruption, the gap to be addressed

Resolution
The shift from the information deficit model to more direct, critical, rhetorical public engagement models (Grabill & Simmons, 1998) has spurred a considerable number of on-the-ground studies and projects with residents in vulnerable areas, often taking form through ethnographies, case studies, interviews, and surveys (e.g., Covi & Kain, 2016; DeLorme et al., 2018).

7.1
Introduction of a Scholarly Article
Source: Stephens and Richards (2020)

technologies. But such technologies, they argue, can still leave the public engaging with data, as in the information deficit model:

> advances in public-facing risk communication technologies cannot leave behind the advances in our understanding of the role of emotions, public perceptions, community, and story. The excitement and availability of profound technological advances should not preclude a situatedness in the lives and livelihoods of those most at risk.

The authors then move into their *niche*, the focus of the article:

> a project that demonstrates how narratives—specifically experiential stories from those in vulnerable regions—can be combined with map-based sea level rise (SLR) risk visualizations to create an interactive, visual tool that gives more context or nuance to the risk.

Methods

Following the introduction, the methods section of an article can be brief, especially if the research used a standard method:

- Procedures followed and materials used
- Applicable assumptions and theory

Because the story mapping article presents the development of a method, an interactive tool, as its main news, it devotes attention to this feature, as the abstract indicates. The final paragraph of the introduction, which previews the article's structure and content, suggests this emphasis on theory and methods:

> This article begins by positioning interactive SLR visualizations as part of the larger communication genre of interactive risk maps. We then discuss narrative and its role in communication, in part as a follow-up critique of the limitations of what we see in many technocratic interactive risk maps. From there, we then describe the origin and development of a story map project focusing on two regions along the east coast of the United States. We describe in detail the rhetorical design decisions (by which we mean the choices we made within the available means—affordances and constraints—of the story mapping application) in constructing a story map that combines the data exploration capabilities of an interactive risk map with visual stories of residents located on the map. We end with a discussion of the potential results and further directions of this approach of combining data with narrative on SLR risk maps.

Results and Discussion

As promised in the introduction, in this segment, you "occupy the niche" by including such information as (Harmon & Gross, 1996, p. 70):

- Experimental or calculated results in text, tables, figures
- Comparison of results (present vs. published earlier, baseline vs. altered state, experimental vs. control group, theoretical calculations vs. experimental measurements)

- Reference to previous research for purposes of criticism or support
- Interpretation of significance of results and comparisons

The discussion section of the story mapping article, much shorter than the methods section, begins with this sentence: "In this section, we discuss the larger implications for communication design and technical communication of the decisions we made when designing the risk map and conducting and editing the interviews." In the discussion, the authors detail the advantages and limits of the template they used, along with the challenges they faced in placing resident stories within the "grand narrative of climate change," and "curating" and editing the stories themselves, always with attention to accuracy.

Conclusion

In the conclusion of the article, aimed at colleagues rather than management, you may include (Harmon & Gross, 1996, p. 70):

- Explanations for surprising or contradictory results
- Main claims derived from having occupied the niche
- Wider significance of those claims to research territory
- Suggestions on future work to validate or expand upon claims

The conclusion of the story mapping article recaps main points detailed earlier, questions why more hasn't been done to alert residents to the problems of SLR mitigation, reiterates the importance of engaging communities in collaborative action, and discusses next steps in the project.

Persona and Style

In a technical or scientific article, the traditional persona has been that of an objective, invisible reporter. The following guidelines provide the generally accepted approach to creating such a persona:

- Be open with information. Detail your method so that others can verify your results by reproducing your work or show why your results are false.
- Use technical terminology to achieve precision and accuracy.
- Avoid emotion, humor, or irony.
- Write in a style that focuses the reader's attention on the object under study and does not draw attention to itself or to you.

These guidelines still hold. But writing about science and technology is increasingly being recognized as a form of argument. Writers and readers establish the truth through discussion. Within the bounds of accurate observation and honest reporting, scientists are less interested in facts in themselves than in facts as illustrations that support some interpretation. Our understanding of natural phenomena thus advances through the collaboration of people who see things differently, take responsibility for what they see, and reveal their personal points of view. Through discussion and publication, professionals convince each other about what is true

For more about persuasion, see Chapter 8

and uncover error. As one authority notes, "The physics of undergraduate textbooks is 90 percent true; the content of the primary research journals of physics is 90 percent false" (Ziman, 1978, p. 40).

Recognizing the argumentative function of reports and articles, many current authors deliberately create a persona that is not "invisible" but suggests trustworthiness by displaying some of the writer's personality and engaging the reader's attention. They avoid impersonal constructions such as "It has been decided that" or "It has been shown that" or "It has been observed that." Such expressions dilute responsibility because they don't say who decided, or showed, or observed. In addition, they make the writer sound stuffy and imply that the reader should simply yield to whatever unnamed authority the writer brings up. An engaging voice in articles, as in all writing, attracts readers' attention and creates a more lasting impression of your work, and of you.

Stephens and Richards use the personal pronoun "we" and speak in a voice that is authoritative but hardly stuffy. They use technical terms familiar to their audience of colleagues. Their relatively long and dense sentences and packed paragraphs also match normal style for a journal article. They include explicit statements that frame the article to alert readers to what's coming, in what order, as in the paragraph at the end of their introduction. Such devices ease reading and remembering.

This emerging voice in American publications strikes a Russian translator as significant. She says that American scientific texts "give a more vivid picture of the problem under consideration, the result obtained or the theories developed" than articles written by their Russian counterparts, which are "academic," and "frozen" in a narrow, unemotional style. "The number of non-neutral, expressive, emotive elements in American texts is large . . . which testifies to the individuality of both the author and of the style of scientific sublanguage in general" (Serebryakova, 1993).

But too much personality in a document may make a reader suspicious. You have to choose how much of yourself to show, depending on your reader's expectations and the type of document you are writing. Too much informality in a scientific article may make readers from another culture suspicious of the author's credibility and authority. For such audiences, you may have to speak not only in another language, but also in another voice, keeping your personality under wraps and inviting less interaction.

Adapting

For more about signaling a reader, see Chapter 10

The next time you read a technical article in your research, pause to take a second look at its structure. See if the content you read in the article fits the categories of information and arrangement you've just learned. Be alert to framing statements and transitional sentences and paragraphs that ease reading. When you come to write your own document, imitate or adapt this pattern to frame your own content.

THE REVIEW OF LITERATURE

Within an article or report, as in the story mapping article, or sometimes as an entire article, you're likely at some point to prepare a literature review. The goal of the review is to "give order to the past, so as to establish a shared present that will be the basis for coordinated work in the future" (Bazerman & Paradis, 1991, p. 5). A report, proposal, or article on empirical research may open with a brief review that:

- Shows how your project fits with other projects and doesn't duplicate them
- Highlights what's new in your approach
- Helps confirm the authority and validity of your work.

Writing a review is a good way for you to master the literature in your field, and thus an instructor may ask you to prepare an entire document that reviews information from a variety of sources and provides a coherent explanation of one issue, technical approach, event, or topic. You'll find examples of such reviews in a variety of publications, but especially in journals devoted exclusively to them, such as *Chemical Reviews* or *Nutrition Reviews*. The *ACS Style Guide* defines such a review as follows:

> Reviews integrate, correlate, and evaluate results from published literature on a particular subject. They seldom report new experimental findings. Effective review articles have a well-defined theme, are usually critical, and should present novel theoretical interpretations. Ordinarily they do not give experimental details, but in special cases (as when a technique is of central interest) experimental procedures may be included. An important function of reviews is to serve as a guide to the original literature; for this reason accuracy and completeness of references cited are essential.
>
> (Coghill & Garson, 2006, pp. 19–20)

In a review article, as the definition you just read suggests, you provide a complete and authoritative account that describes which researchers are working on the topic and with what success. Because engineers, for example, are legally responsible for knowing all relevant information in a field when they select a material or complete a design, such reviews provide a valuable service. Reviews are also useful in drawing together information from various disciplines—economics, law, engineering, science, psychology—to focus on problems that necessitate broad understanding, such as the design of earthquake-resistant highway overpasses or the implementation of recycling programs. But completeness in a review is not clutter. Be selective about what you display as evidence. Discard irrelevant information, no matter how hard it was to come by, how interesting it may seem, or how clever a researcher it proves you to be.

Structure

The writing of review articles presents special problems in fitting together materials from disparate sources so they form a coherent presentation and not a simple collage or mood board. You develop coherence by establishing a theme and then framing the content to support it. The theme in the article on story mapping, for example, is the shift from an "information deficit" model to a "public engagement model." Within a general theme, two frameworks are common and sometimes interwoven:

1. Show how authorities either agree or disagree on key issues and findings.
2. Move publication by publication, in chronological or logical order.

The following paragraph shows how authors agree about the importance of the physical environment in researching how people collaborate in a workplace: Note the integration of direct quotations into the review:

> Gieryn (2000, 2002) addresses ground rules and conditions for keeping an eye on the physical environment as it impacts issues of interest to sociologists of work. Organizational studies scholars, for example, Nicolini, Mengis, and Swan (2012), also emphasize the need to examine such workplace behaviors as collaboration in practice, not just from a broad humanistic perspective. One approach is to look at the "material infrastructure" that sustains it, including the spaces and furnishings in which it occurs (p. 625). Others concur. For example, Olsen (2013) argues that "our active cohabitation with things regulates and routinizes our behavior, making it repetitive and recognizable" (p. 181). Objects have the capacity to instruct us in learning to behave and in developing the skills needed to exploit their affordances, as they also foster a sense of community and belonging.

The author of the next paragraph frames the discussion in terms of the disagreement of authorities concerning the role of craftsmen in technological advances. Citations are pegged to a numbered list of references at the end of the article:

> Melvin and Robinson [19] argue against the notion that most technological advances were the result of the work of anonymous "little men," craftsmen and technicians often innocent of basic scientific development and training. They think that to insist upon the importance of such craftsmen "serves to reinforce the belief in the autonomy and purity of science as a concept-generating activity. . . . Ruling out any significant interaction between technologists and scientists until the late nineteenth century [merely] protects a particular historiographic viewpoint," a viewpoint that has tied to it "a particular kind of scientific community" (p. 78). Their opinion is in direct contrast to that of Thomas Kuhn [24], who sees the polarization between technology and science as springing from subterranean roots, "for almost no historical society has managed successfully to nurture both at the same time" (p. 50).

Style

These model paragraphs demonstrate successful strategies for achieving a unified voice in a review. Three matters of expression that require attention in any writing project are particularly significant in reviews: integrating quotations, selecting the appropriate verb tense, and citing references.

Integrating Quotations

The quotations in the two paragraphs you just read highlight key terms and succinctly state an important argument. Overused, however, quotations impede the flow of the text. They indicate an author who, out of laziness or modesty, failed to tailor sources to the new occasion of the review. As an author, you need to rework others' material into a new context, for a new reader, so avoid long quotations. If you feel compelled to quote extensively, determine the best style in your communication product for indicating quoted matter and implement that style consistently.

Selecting the Appropriate Tense

Adhere to the common conventions of the review when you select verb tenses. Those conventions dictate that you use the present tense to discuss authors whose ideas you consider valid or current:

> Smith (2019) argues that wetland restrictions are ineffective. His complaints are also validated by Rogers (2020).

Use the past tense to argue for the lack of currency in an interpretation that has been corrected. Use the past, too, to indicate the real time in which a case study was conducted or a test performed:

> Smith's criticism is based on a 2015 study in which he surveyed 25 federally designated wetlands in the Delaware Valley.

Citing Sources

Provide both enough citations—that is, include all the major works—and complete citations—that is, adequate bibliographic information so the reader can find the original source. In addition, tuck that information into the text as noninvasively as possible so the reader isn't thwarted in getting the gist of the review because of an intrusive referencing system.

This chapter has suggested strategies for finding, evaluating, and incorporating secondary sources effectively as well as ethically in your research and in your writing. These sources are available both internally, where you work or study, and externally, in print, on the web, and in social media. Select and mix insights and information from sources to reuse in your own communication products. Through that process, you add value to what others have said on the topic you're addressing. Be sure to acknowledge your sources through appropriate and conventional documentation. In addition, use your reading of technical articles as a method for learning to compose such articles yourself, adapting their structure and style. The articles can also serve as models for effective argumentation in other genres.

CHECKLIST: INCORPORATING SOURCES: FROM RESEARCH TO COMMUNICATING

1. Read secondary sources to inform yourself about what others have learned and claimed about a topic
2. On the job, consult sources that circulate within your company and to its stakeholders
3. More broadly, consult the vast quantity of *external* sources existing in three large spheres
    Examine technical and scientific journals for the most reliable information
    Look into the public sphere of information available on social media
    Be aware of how the algorithms of the platforms on which you read bias your reading

4. Filter, reuse, and integrate content from sources into your own document
    Use direct quotation sparingly but effectively
    Paraphrase a source to condense or highlight
    Recognize the ownership of words, ideas, images, and points of view
    Establish your own authority by citing experts
    Apply the rules and conventions for appropriate documentation

5. Adapt the model of the technical article for your own writing
    Consider a structure of introduction, middle, and ending
    Create the appropriate persona to establish trust, including self-revelation

6. Write a review to integrate, correlate, and evaluate results from the literature on a topic
    Establish a theme
    Trace the chronology of sources on that theme
    Support that theme by saying how authors disagree or agree
    Integrate quotations
    Use appropriate tense
    Provide adequate citations of major sources
    Provide complete documentation for those sources

EXERCISES

1. Write a memo addressed to students in your major ranking the five most significant journals in your field. For each journal, include the following information:

    - Complete citation
    - Circulation (number of subscribers) and impact statistics
    - Coverage: the general topics of the articles
    - Reliability: how do you know you can trust its articles?

    Justify your choice of the journals as "significant."

2. The question of reliability arises in particular when you deal with information circulating online. You're faced with both misinformation and disinformation. The chapter cited some of the many sources available to help you sort the reliable from the fake. Your campus library also probably provides online guidance. Here are a few additional links to help you evaluate online sources, especially social media. Use your research skills to uncover others:

   - From the University of Delaware Library: https://bit.ly/3vshjga
   - The Public Editor Project: https://bit.ly/2R29RJU

3. Find two articles, one from a technical journal and one from a source oriented to a wider audience, that discuss roughly the same topic—for example, the incidence of concussions among players of certain sports, the effects of legalizing marijuana in many US states, the decline in the number of bird species, or the relative safety of various electric or autonomous vehicles. Compare the articles in the following categories:

   - Audience (how do you know who reads these articles?)
   - Format (conventional segments or free-form, obvious or subtle design)
   - Style (length of sentences and paragraphs, use of technical terms, general readability)
   - Content (theoretical? practical? supporting evidence—examples, stages in a logical argument, statistics, interest-getting devices?)
   - Visuals (number, type, purpose, style)
   - Delivery (were the articles enhanced though options in multimodal digital delivery?)

   Then write a memo addressed to fellow students comparing the communication strategies used by the authors to address the needs of their intended audiences. Include links to the articles.

4. Write a brief review of literature (literature in a very broad sense) on a topic of interest. You might continue the investigation you conducted in Exercise 3. Include four sources. They may either all support an argument in your sources or contradict each other.

FOR COLLABORATION

In your health-and-nature team (collaborative exercise for Chapter 6), turn now to using secondary sources. Where would you look for such sources? Which external sources would be most appropriate for gathering information to advance your investigation of the particular problem your project aims to solve? Create a list of five to ten sources, briefly look at each, and then suggest how each might contribute to your research. Share the list in a discussion with members of the other project teams in the class.

REFERENCES

Applen, J.D. (2020). Using Bayesian induction methods in risk assessment and communication. *Communication Design Quarterly*, *8*(2), 6–15.

Bazerman, C., & Paradis, J. (1991). *Textual dynamics of the professions*. Madison, WI: University of Wisconsin Press.

Beckerman, G. (2018). Kicking the geeks where it hurts. *New York Times Book Review*. June 10, p. 14.

Coghill, A., & Garson, L. (Eds.) (2006). *The ACS style guide: Effective communication of scientific information*. Washington, D.C. American Chemical Society.

Dusenberry, L., Hutter, L., & Robinson, J. (2015). Filter. Remix. Make: Cultivating adaptability through multimodality. *Journal of Technical Writing and Communication*, *45*(3), 299–322.

Gilbert, S. (2020). In a pandemic, what use is Google? Methods, innovation, skills. Sage Ocean. June 9. Retrieved from https://bit.ly/2RGeTvR

Harmon, J., & Gross, A. (1996). The scientific style manual: A reliable guide to practice? *Technical Communication*, *43*(1), 61–72.

Lam, C., & Biggerstaff, E. (2019) Finding stories in the threads: Can technical communication students leverage user-generated content to gain subject-matter familiarity? *IEEE Transactions on Professional Communication*, *62*(4), 334–350.

Leijten, M., Van Waes, L., Shriver, K., & Hayes, J.R. (2014). Writing in the workplace: Constructing documents using multiple digital sources. *Journal of Writing Research*, *5*(3), 285–337.

News Literacy Project. (2021). How to know what to trust. Retrieved from newslit.org

Serebryakova, I. (1993). Translation problem of English metaphoric terms into Russian. Unpublished.

Stephens, S., & Richards, D. (2020). Story mapping and sea level rise: Listening to global risks at street level. *Communication Design Quarterly*, *8*(1), 5–18.

Tingley, K. (2020). What does it mean for science—and public health—that scientific journals are now publishing coronavirus research at warp speed? *New York Times Magazine*. March 26, p. 18.

Ziman, J.M. (1978). *Reliable knowledge. An exploration of the grounds for belief in science*. Cambridge, UK: Cambridge University Press.

Part 3
Designing Content for Audiences

8 Explaining and Persuading

This chapter begins a series of chapters focused on developing and designing content for the various audiences you'll address as a technical professional. In a broad way, think about the communication purposes you share with your audience as ranging from explaining to persuading. In an explanation, you help your audience to understand some concept, object, action, place, or system. You also enable your audience to do something, such as filling in a form. Your explanation may play a role in achieving another purpose: persuading your audience to accept an idea you think is valid or to act in the way you intend.

In this chapter and the chapters that follow, you'll learn detailed, mix-and-match strategies for using text and visuals to explain and persuade. This chapter focuses mainly on text. Chapter 9 focuses mainly on visuals. In the process, you'll make complex information accessible to those who need it, encourage action and remembering, and create communication products that look good and make sense as they make you look good, too.

EXPLAINING

To compose an explanation—that is, to turn information you've collected or know into content for an audience—call on one or more of three well-tested techniques: describing, classifying, and defining.

Describing

To describe what something looks like or how it operates, you'll probably start with a picture: a photograph, drawing, computer-aided rendering, or other digital image such as those you'll learn about in Chapter 9. Is that alone enough for your intended audience? The answer is often *yes*. Think of the cartoon-like instructions that accompany flat-packed furniture. Those instructions easily cross borders of language and skill. But you may add text to amplify the explanation for an audience unfamiliar with what you are describing.

Describing an Object or Mechanism

To see how pictures and text work together in an explanation, take a minute to look, for example, at the business plan reproduced in Chapter 14. Under "product description," the authors explain their innovative idea for accommodating adaptive

DOI: 10.4324/9781003093763-11

users of a rowing shell. They include four visuals. One is a computer-aided rendering to help readers understand the prototype propulsion system. Two photographs of rowers in action further explain the system: one rower uses a standard system, and another uses the new system. A third photograph depicts the first-generation prototype of the system, which was used for validation and proof-of-concept in controlled, water tank testing.

You may devote a segment of a document to the description or an entire document, depending on your purpose and the importance of the object or mechanism. The box on "General Plan for the Detailed Description of an Object" provides an outline for an extended description, to be adjusted to fit your purpose in a particular situation.

GENERAL PLAN FOR THE DETAILED DESCRIPTION OF AN OBJECT

1. Introduction. An overview

    - Definition
    - Purpose/benefits
    - General principle of operation
    - General appearance (often a visual)
    - Division into components

2. Component-by-component analysis

    - Purpose of components
    - Appearance
    - Details: shape, size, relationship to other parts, materials, color, and so on

3. Conclusion

    - Special point/advantages (or disadvantages)

Describing a Process

In design-thinking-oriented projects, you'll often find yourself describing processes, a series of steps aimed at some goal. One genre of process description is a *specification*. Specifications include, for example, standards for testing and measuring the quality of a product and methods for building, installing, or writing something. The result of your investigation of a problem of fact or means may consist of a description of a process. In the description, you may take the point of view of how something happens (how a virus enters the human body, how it is transmitted from body to body), or how to do something (how to set up a drive-in testing facility to determine who might be positive for the virus). Your explanation may include visuals, but you'll need to provide introductory and concluding statements to help the audience understand its context. Because of its importance, explaining how to do things in a set of instructions is discussed at length in Chapter 17.

For more about problem categories, see Chapter 5

For more about
describing a method,
see Chapters 15
and 16

Another term for the description of a process is a *narrative.* That term frequently designates the section of an article or a report in which you describe what
you did to solve a problem or create a product—that is, your method.

Another form of narrative is a story. Sometimes thought of simply as fiction
or personal storytelling, narratives are becoming increasingly important as a way
for health care professionals, corporations, and entrepreneurs to engage with
their clients and customers. The framework of a story can embed messages and
draw attention. Successful entrepreneurs create compelling stories or scenarios
about how an innovation they are producing can enhance the future lives of their
customers.

For more about
storytelling, see
Chapters 14
and 19

Classifying

For more about
tables, see Chapter 9

Anther strategy for explaining facts as well as concepts is to classify them. You
identify common elements in natural and abstract phenomena to help your audience (and you) understand what things are and what makes things run. Choose a
principle of selection (color, size, organizational type), give it a key word, and then
put items into categories. The most vivid form for showing classification is a table.

How you classify items may reflect cultural values. For example, a researcher
showed Chinese and North American participants in a survey three pictures—a
cow, a chicken, and hay—and asked them to group two objects together. The Chinese chose the cow and hay (the cow eats hay) in terms of relationship, and the
North Americans chose the chicken and cow (as animals) in terms of their attributes
(Page, 2003, p. 141).

Defining

A third strategy for explaining, for helping an audience understand your information,
is defining the terms you use to represent your subject. Defining terms is a critical
phase in design-thinking projects, as you learned in Chapter 5. It is critical, too, in
presenting the results of those projects. You and your audience need to agree on
major terms in your discussion. While visuals may accompany a definition, defining
as a strategy for explaining ultimately comes down to words.

What is *sustainable agriculture*? Define each term to help the reader understand the concept. If your reader is familiar with the idea of agriculture in general,
but not with the idea of *sustainability*, focus on the new practice within the context
of what the reader will find familiar. The extent of your definition and its location in
your communication depend on the reader's degree of familiarity with the concept
or item and the purpose that you and the reader share. Sometimes, a simple phrase
is enough. Saying, for example, that *compression* makes things shorter and *tension*
makes things longer may define those terms adequately for a popular audience, if
not for engineers.

The general pattern for a formal definition is to classify the term into a category
and then show how it differs from other items in the class. For example:

Flood stage is the *height of a river* above which damage starts, typically because
the river overflows its banks.
Diet is the *food or drink* normally consumed by an individual or a group.

Be sure to accommodate the audience by using terms that are more familiar and clearer than the term being defined and by not repeating the term in the definition. Your agreement with teammates and your audience on the definition of key terms may need to be stated explicitly in the introduction to a report to limit expectations about what will be covered in this context. For example:

> In this report, the term *competencies* refers to the knowledge or skills that an employee has or needs. We focus on three competencies: technical, business, and relational.

In defining concepts, adjust how much of your attitude you let show. Especially when you use comparisons as part of an explanation, deal with controversial topics, or are explaining something to someone in another culture, explicitly check for your own biases and point of view. For example, look at these two definitions of clearcutting, a forestry practice, from opposing perspectives:

> *Paper company*: True silviculture clearcutting—the removal of an entire stand of trees in one cutting—is a useful forest management practice that can promote the grown of a high-quality forest. Clearcutting has both short-term and long-term benefits as part of an overall forest management plan.
> *Environmentalist*: Clearcutting is simply a hemorrhaging of the forest. Paper companies shave a mountaintop down to the ground; the denuded mountainside is almost certain to erode during the next rain.

To some degree, that "true" in the paper company's statement suggests the company's recognition of the opposition. A presentation based on a biased definition of a key term is sure to further misrepresent the topic and mislead the reader. This occurs, for example, when producers discuss the importance of coal-fired energy by introducing the concept of "clean coal." They are arguing that a carbon-based fuel is consistent with a sustainable environment, not a concept generally agreed upon. In addition, watch for unintended side effects when you use metaphors (such as "hemorrhaging")—that is, when you give something an attribute that belongs to something else. As with any writing, be aware of the connotations, the associated meanings, of the terms you use when you explain something and when you write across cultures (see the box on defining terms across cultures).

DEFINING TERMS ACROSS CULTURES

Language carries values and perspectives, a lens through which to view the world. For example, Christopher Newell (quoted in Voltin, 2020, p. 22), a member of the Wabanaki Nation in the US, notes:

> In Passamaquoddy, we have a term, *tupqan*, which we translate for English as "dirt, soil, or earth." But the more literal translation is the "molecules of

our ancestors." Through the lens of Passamaquoddy, the dirt is not something destined to be improved or developed by humans, but rather a living animate force of the natural life cycle. It shows a highly scientific understanding of the local eco-system. Because of this, the land (as a living force) cannot be owned as individual property. Our language was born of the land and only truly makes sense in that context.

PERSUADING

When you expand or limit the common definition of a term, or when you move from factual descriptions of what a mechanism looks like to asserting that your new mechanism works better than others in a situation of interest, you enter into territory where people may disagree. Your first task in communicating is to assess the likelihood that you and your audience have enough in common that they will simply accept what you say as true. If not, you need to persuade them.

Three Strategies for Persuasion

Broadly understood, to persuade, you adjust your ideas to the audience and the audience to your ideas. Many years ago, the Greek philosopher Aristotle identified three strategies for persuasion. One is *logos*, or the logical appeal based on providing good reasons. Second is *ethos*, or the appeal to the character and authority of the speaker or writer. Third is *pathos*, the emotional or aesthetic appeal. Updated and enhanced through current technologies and design thinking, these appeals still have relevance today. They can be effectively integrated into your strategies for persuasion. The rest of this chapter is divided into sections that provide advice on each of the three persuasive strategies.

Constraints on Persuasion

Before implementing these strategies, however, be sure to recognize any potential constraints in the situation you face. A key constraint, for example, is timing. Determine the best time to deliver your message by assessing the message, the audience, and the environment.

Message

Is your message ready? That is, do you know the outcome you want, and the outcome your audience wants? Do you have all necessary information, in the right arrangement? Are you confident about what you have to say?

Audience

Is this a time when the audience can attend to the message and is not, for example, distracted or too busy? Are they focused on the issue—or inclined to be focused? Do they consider this issue important? Is the audience *able* to act? Can they do what you want?

Organizational or Community Environment

Does your message fit the culture of the organization at this moment? Does it build on other ideas or messages circulating in the organization now? Does it conflict with other messages? Does prevailing corporate policy or organizational culture mean certain failure if you pursue the idea in your message?

In general, control what you can. Make sure your message is ready before you attempt to persuade someone else about its validity. Assess the situation to determine if the level of interest in the problem you want to solve is sufficient to warrant an audience's attention. Have recent storms and flooding, for example, made a community responsive to hearing a message about mitigating the effects of sea level rise? Find a way to highlight the priority of your topic. If you have a cost-saving proposal, present it when your manager is reviewing the budget.

Relate your interests to those of your audience. Don't simply assume they share your enthusiasm. How will your idea solve a problem they have? Is your audience ready, psychologically and developmentally, to behave in the way you think they ought to behave?

Throughout the process of persuading, engage your instinct for community. Don't lie or stretch the truth to the detriment of others. Leave others at least as well off as you found them. If possible, leave them better off. Be inclusive and empathetic.

For more on ethical communication, see Chapter 3

PROVIDING GOOD REASONS

In making a logical appeal, you provide good reasons to prove you're right. You support a new or previously uncertain idea with information that, at least at the time, is certain, easily verifiable, and could generally be agreed upon. You also make sure to select information your audience will find compelling. Providing good reasons is a complex process, well codified in traditional principles of logic. But here, briefly, are two strategies to consider as you develop and support your own assertions: thinking by analogy and analyzing causes and effects.

Analogies

Through analogy, you try to understand how elements of a new phenomenon relate to one another in a way similar to how elements of known phenomena relate. Analogies help you understand an idea and help your audience understand and then accept that idea. Analogies are effective when your audience finds the similarities relevant and significant and can't point out any dissimilarity that is even more important.

A key analogy, for example, is at the core of design thinking: By building prototypes and models rapidly and cheaply, you can anticipate and correct problems in design early to ensure things will work when you scale up to full production. Scientific researchers, too, operate by analogy in thinking that what they demonstrate in the laboratory will be replicated in a similar population or situation in the world at large. An example is research conducted on hamsters to test the efficacy of masks to prevent the transmission of viruses among humans.

Thinking by analogy also helps you develop new theories and techniques. For example, sociologists have developed the theory of "social contagion" to explain similarities between the way ideas spread among populations and the way viruses spread among them. Later in the chapter, you'll read about how that concept helped shape communications aimed at persuading students to behave appropriately in response to the COVID-19 virus. Researchers also used an analogy—to Swiss cheese—as a way to explain the control of natural hazards in society (see box).

THE SWISS CHEESE MODEL

To control the hazards that come with today's complex interactions of human and natural elements requires layers of defense and intervention. Ideally, each layer is intact. But researchers discovered that controlling the COVID-19 virus suggested more of a "Swiss cheese model." Each layer, like each slice of Swiss cheese, has many holes, which wouldn't necessarily cause problems. Only when the holes in many layers line up do hazards come into damaging contact with victims (Christakis, 2020, p. C1).

Cause and Effect

Analogies help you understand. But they can't prove anything. Proving brings in another logical dimension. Often, it's about matters of cause and effect. Given an event or phenomenon in the past, can you identify what caused it? Can you, on the basis of that information, predict a future event? These questions are critical to much technical and scientific research. What may happen is that you mistake a mere correlation in your information for a causal relationship. One thing following another over time or two things accompanying each other does not necessarily mean that what followed was the result of what happened earlier (see box).

You need to sort through the necessary causes that must be present for the effect to occur and the contributory causes that may also be in the picture. For example, observers may conclude that low wages are a necessary cause of low morale in an organization. But the wages may be only a contributory cause, along with other causes, such as the lack of safety precautions, inadequate health benefits, or a poor physical environment for work.

SUPPORTING AN ASSERTION WITH CAUSE-AND-EFFECT REASONING

The harvest of fish off the coast of New England declined 62 percent between 1982 and 1993. The fishery collapsed. Why? One researcher asserted that the US federal government caused the collapse. In the 1970s, the government provided loans and tax credits that motivated investors to outfit a large fleet of big boats with high-tech gear; that fleet fished the waters beyond their ability to regenerate. Many fishermen, especially those who own small boats, would argue that the buildup of big boats was indeed a necessary cause of the fishery collapse. But contributory causes including pollution, the destruction of some fish habitat, and a rise in the number (and appetite) of seals, which prey on fish, may have been more significant. A few scientists also dispute the notion that the fishery did collapse. They see the decline in the mid-1990s as part of a long-term natural cycle.

Uncovering Errors in Reasoning

To argue against someone else's assertion (for example, that the government caused the fishery collapse)—that is, to disprove what they are trying to prove—you point out errors in their reasoning. You provide such contradictory evidence as facts that weren't acknowledged, data that are missing, a bias or lack of randomizing in the sample, or workable alternative explanations. Start your counterargument by looking for problems in causal reasoning. Here are a few other common errors.

Faulty Generalization

You generalize falsely when you leap to a conclusion without adequate evidence. The observations on which you base your statement may be irrelevant, unrepresentative, or not numerous enough. You may oversimplify or fail to incorporate cultural differences.

"Working from home (WFH) is more productive than working in an office." That may be true for some people. Early in the 2020 pandemic, when a large number of companies sent people out of the office to work at home, some managers noted increases in productivity among their staff. But research has also shown that an initial bump in productivity may be short-lived. The stress and anxiety caused by a 24/7 blurring between work and personal life form a counterargument, as is a potential loss in innovative thinking when people don't share a physical space for conversations and collaboration.

Implied Assumption

You may take things for granted in a discussion—for example, that being healthy is better than being sick, and thus people would want to behave in ways that promote health. But sometimes that assumption is false, as in the following examples.

- A picture in the newsletter of a large chemical company showed an executive unrolling before her the 20-foot-long printout of a chart detailing the steps required for one plant to meet the standards of just one federal agency. The implied assumption: If regulations take so much paper, they must be excessive. Someone taking a different perspective might point out, reasonably, that the length of the flow chart of the regulatory process is not necessarily proportionate or even related to its appropriateness.
- "Our team worked 15 hours on this report and got a C. Their team tossed in their first draft and got an A. Your grading isn't fair." This reasoning, a special

form of implied assumption, begs the question about the *quality* of the final product. The grade represents what the work accomplished, not how long the work took.

Either/Or

Your conclusion also doesn't follow the evidence if you see in a situation only one alternative when there are none—or more than two, as in this example:

"To solve our need to conserve energy, we must have innovation and imagination from architects, not legislation from Brussels." This statement sets two alternatives against one another. In practice, energy conservation requires a combination of imagination and enabling government legislation.

Presenting a Logical Argument

Using analogies to derive explanatory statements and using evidence and reason to prove and disprove such statements are at the core of a logical approach to persuasion. A pattern of cause and effect, for example, provides an effective way to organize a persuasive discussion. The following paragraph uses such a pattern to help support a transportation department's appeal to implement an alternative to salt in winter road care:

[effect: potholes]

[cause: salt]

Sodium chloride causes potholes in highways. Because concrete is porous, the salt penetrates the concrete until it reaches the reinforcing steel beneath. When the salt reaches the steel, the steel rusts, and the rust occupies a larger volume of space than the steel from which it came. The expanding rust then causes the concrete to crack and form potholes.

For more about IMRAD reports, see Chapters 7 and 16

You'll use a pattern of assertion and support, of proving and disproving, in many communication products. The box on "A Time-Tested Plan for Presenting a Logical Argument" shows how this pattern can be adapted for an IMRAD report.

8.1
Persuading with
Cause-and-Effect
Reasoning

(Notice attached to windshields of cars in a ski area's parking lot)

Effect: saturated snowpack

After nearly a week of wet weather, the snowpack is saturated with moisture.

Another bad effect: crusty surface

Temperatures dropped unexpectedly fast this morning, producing a crusty surface, particularly at the upper elevations.

Balance of long-term good ("best possible conditions") with short term inconvenience ("delaying opening")

On the lower elevations, this crust is thin enough that is should break up with skier use. However, to ensure the best possible conditions, we are delaying opening this morning for 1/2 hour to permit grooming operations to start. We'll be closing certain trails as the day progresses to permit grooming. You may expect these trails to re-open once they've been groomed.

Goodwill compensation

Because surface conditions this morning are less than what we'd like to offer, we invite you to clip today's ticket after skiing today and bring it back at any time prior to December. This ticket will provide you with a $10 discount on your next day of skiing.

A TIME-TESTED PLAN FOR PRESENTING A LOGICAL ARGUMENT

- Introduction: The problem initiating your research and your major assertion
- Statement of facts: Your supporting evidence, often a narrative of your method
- Confirmation: Your results, your good reasons, including examples
- Refutation: Why any opposition is wrong, because it depends on either false reasoning or false information
- Conclusion: Restatement of the major assertion as true or certain because proven.

ESTABLISHING YOUR GOOD CHARACTER AND INSTINCT FOR COMMUNITY

The approach to presenting a logical argument outlined in the box has more to do with adjusting your audience to your ideas than about adjusting your ideas to them. It minimizes your recognition of the role they might play in coming to a decision or supporting a common idea. But recognizing their role extends the reach of the second persuasive appeal identified by Aristotle: the ethical appeal. Originally centered on the character of the speaker or writer, this appeal is easily extended to encompass a larger ethical concern for your community.

Developing a Trustworthy Persona

Trust in an expert has become an increasingly important strategy for persuasion as data collection expands exponentially with new technologies. In uncertain and rapidly changing conditions, with competing claims about what's true, people look to experts to interpret what to believe and determine what to do. In high-context cultures, such trust, along with shared values, can be enough to encourage agreement and action. But in the low-context situations frequent in today's global economy, even experts can't assume such easy acceptance. They have to prove themselves worthy of trust. They have to establish a persona in their messages that convinces others to trust them. As a communicator, you have to create such a persona, too. Here are suggestions for establishing a persuasive persona in your communications.

For more about high- and low-context, see Chapter 1

For more about persona, see Chapters 7 and 11

Demonstrate the Personal Qualities Your Audience Values

Determine those qualities and balance them against qualities valued in your own community. If you pride yourself on individual assertiveness and directness, you may have to tamp down that approach to appeal to people who find such assertiveness offensive.

Show Your knowledge of Authorities the Audience Respects

Citing experts helps validate your own work, justify your stance, and show that you recognize the collaborative nature of all scientific and technical pursuits. In cultures

For more about citing sources, see Chapter 7

where relationships are highly valued, demonstrating your position in a network of the right people may strengthen your credibility.

Show that You Are Well Informed

The nature and ownership of knowledge vary across cultures. You enhance your credibility as an authoritative source when you recognize where information is needed and produce the right information in a form your audience can understand.

Acknowledge Your Point of View

A long-standing value in Europe and North America is that technical and scientific communication is objective. Increasingly, however, objectivity is being recognized less as something fixed and more as a point of view, among others, in the global community.

For more on objectivity, see Chapter 7

Understanding Cultural Differences

In trying to persuade—that is, to adjust your ideas to others—learn what makes others comfortable and willing to accept ideas and change their behavior. People—perhaps you—often act in their own self-interest, may resist change, and may resent and challenge opposition. You have to frame any change as less threatening than they may think and call on higher values or interests as necessary. You have values and norms for action you accept as true. So does your audience. For example, recognize the reluctance of those whose view of life is fatalistic ("What will be, will be") to forgo their pleasures in smoking or drinking to excess even if you show them their increased risk of cancer or other diseases. Your discussion of long-term consequences won't be persuasive with those who pay little attention to long-term risks. In determining strategies for managing a forest, an environmentalist may emphasize the need to preserve endangered species of wildlife. Farmers and loggers, who have a different perspective, may see the land as an economic resource.

Differences in perspective may reflect your training in a discipline, such as environmentalism, or your work or economic interest, such as farming or logging. In addition, they often reflect cultural differences. Generalizations about culture are risky, and the global marketplace is putting many traditional values to the test. But differences in perspective and values remain and need to be understood. This section briefly suggests some dimensions of difference in the global community.

Confrontation versus Harmony

In the US and the UK, for example, low-context cultures, the kind of logical approach outlined in the box on "A Time-Tested Plan for Presenting a Logical Argument" is not only acceptable but desirable in many professional communities, especially legal ones. People debate issues openly in courts of law and meeting rooms. That's how individuals distinguish themselves. The goal is winning, and the process is often described in military terms: tactics, opponents, a war of words, battle lines, the global arena. An investigation of an opponent centers more on finding weaknesses and vulnerabilities than on understanding them. In addition, the argument can take on a life of its own, separate from the arguers.

But in valuing harmony, the Chinese have traditionally seen individuals as part of a collective where debate and open discord are not welcome. The argument can't be separated from the people involved. In such high-context settings, decisions may be made more on intuition, on prevailing moods, and on a need to keep harmony. The message resides in the situation and in the people, less as an objective abstraction.

The Japanese language, for instance, can be less precise and pointed than English, more tolerant of ambiguity. Open-ended or obscure statements provide a kind of buffer zone that avoids personal attack. Such indirection helps save face for everyone. To see how those differences in persuasion can be misunderstood, consider a consulting firm that trains professionals in international negotiation. It says that it can help Europeans "get behind the inscrutable mask" of their Japanese colleagues. Thinking that ambiguity is simply a mask rather than an embedded value dooms the approach.

Individual versus Group

In addition, Japanese organizations traditionally arrive at decisions and settle issues collaboratively. Proposals for action are circulated throughout the organization to gain consent in advance—to build a context in which the decision then becomes inevitable. In traditional US companies, an executive often imposes a decision made individually. The emphasis is also on efficiency and individual performance. The US system tends to be "fast–slow–slow–slow": a quick decision may take some time to implement. The Japanese is "slow–slow–slow–fast," because consent is achieved in the process so that implementation is speedy.

Such differences in values between individual efficiency in corporations and group relationships emerged explicitly in research with a sample of American, Japanese, and European middle managers. They were asked which statement of two was more correct: (1.) "A company is a system designed to perform functions and tasks in an efficient way. People are hired to fulfill those functions with the help of machines and other equipment. They are paid for the tasks they perform." (2.) "A company is a group of people working together. The people have social relations with other people and with the organization. The functioning is dependent on these relations." About 75 percent of Americans chose (1.); about 66 percent of Japanese chose (2.); Europeans were split roughly 50:50 (Page, 2003, p. 140).

Developing Common Ground

Understanding someone else's perspective, their values and norms, can help you develop a collaborative strategy to create common ground. What you create together may well be better than what either you or the others could have achieved alone.

For more about creating common ground, see Chapter 3

Start by establishing the right situation:

- The right timing
- The right arrangement of content and media for delivery
- The right people

First, even the best message sent at the wrong time—when the audience is unprepared to accept it—will fail, as you learned earlier in this chapter. Pay attention to timing. Second, construct the right arrangement of your argument. In doing so, put aside a merely defensive argument such as that outlined in the box on "A Time-Tested Plan for Presenting a Logical Argument": "I'm right and here's why." Think in terms of "we," not "you," as a way to reinforce the relevance of the strategy to all parties. No one is being left out. Rather than simply seeking closure in an argument, invite the audience to agree. The box on "A Collaborative Plan for a Document Demonstrating Common Ground" suggests how you might arrange such a persuasive presentation. Significant, too, is delivering it in the right mode and media as you explain and support your central assertion and persuade the audience. Third, select the right people, in relationship with you, to receive your message.

A COLLABORATIVE PLAN FOR A DOCUMENT DEMONSTRATING COMMON GROUND

- Introduction: Restatement of the problem and the situation
- Statement of valid points in the other's position
- Statement of your position
- Demonstration of common ground.

In the introduction, you restate the problem and situation in a way that recognizes your understanding of the other's position. The next segment describes a context or situation in which their position may be valid. Following that discussion, you state your position, positives first and then any possible downsides, along with evidence drawn from the contexts and situations in which your position is valid. Finally, you demonstrate the common ground between those positions. You focus on how the other's position would benefit if they were to adopt elements of your position and perhaps how you've changed your mind to reflect your understanding of their need. The best outcome is an ability to show how each of you supplies what the other lacks. This strategy usually derives from extensive discussions with the audience in advance of any final document and represents your joint thinking (Young et al., 1970).

Appealing to a Sense of Community

Your ability to persuade an audience depends significantly on your ability to create common ground with them. Those who don't see an immediate benefit to themselves may resist your advice. This need for common ground was demonstrated in multiple ways as governments tried to persuade citizens to behave well during the COVID-19 pandemic. One major strategy, in line with Aristotle's ethical appeal, was to turn to experts. Trusted government authorities, health care professionals, scientists, and doctors took to the media to encourage people to do what seemed, at least at the time, right, such as using, or not using, masks.

Relying on Experts

Turning to experts worked in many cultures, but the strategy was often challenged by individuals in the US. In an opinion article, for example, even two medical scientists called this strategy "paternalistic and outdated." "Protecting people from doubt is central to the strategy," the authors charged. The particular issue being debated was how to assure citizens that a vaccine was safe. Instead, the authors suggested that experts should "assume you're speaking to mature, self-interested decision makers, offer transparent and comprehensive information about the risks and benefits of the vaccine, and engage patients in the decision-making process (Kaplan & Frosch, 2020). The authors themselves, however, it is worth noting, demonstrated some implied bias by calling these decision-makers "patients."

Offering "comprehensive information" is in line with what has been called an "information deficit" model for communicating information about risks. It assumes that, given the right information, people will be persuaded to act in the right way. But that implied assumption may be false. Scientific information, particularly in such uncertain problem areas as novel viruses and untested vaccines, is nuanced, subject to conflicting assertions even within the scientific and health-policy communities, and difficult to interpret. The information may overload users.

For more about this model, see Chapter 7

Blogs and other popular communication genres that face outward from experts to popular audiences can help to stimulate interest and provide reasons for people to behave appropriately. In writing for these audiences, however, experts confront "the dilemma of incomprehensible accuracy or comprehensible inaccuracy and the fun of [the] work lies mainly in the solution of that problem" (Grant & Fisher, 1971). It's important that any inaccuracy doesn't interfere with a reader's ability to gain additional information on the topic. The definition of "compression" you read earlier serves such a purpose for a general audience. That scientists and communication experts might have "fun" in finding the proper balance between inaccuracy and accuracy is an attitude you might assume when you confront the dilemma in your own communication products.

Relying on the Community

Simply confronting the dilemma is one way to create common ground. Experts helped during the pandemic in 2020. But another approach proved more effective in some circumstances. The assumption underlying the approach, aligned with the theory of *social contagion*, is that we are more likely to engage in "prosocial behavior" if we think many others are doing so as well (Tingley, 2020, p. 14). Seeing others in our community behave in a certain way, such as wearing masks, leads to imitation. It clues us into the right response to the situation and relieves anxiety about what we should be doing. Given this research, authorities who tried to persuade by depicting risky behavior, such as groups of unmasked students holding parties and ignoring guidelines, unintentionally suggested such gatherings as a norm and only reinforced that behavior. More effective was creating a new norm by emphasizing those who *were* taking safety precautions, building a norm of following the rules. A persuasive campaign based on statements or visuals about what a majority of the students did well demonstrated improvements in compliance, even when the effect wasn't clearly observable like the wearing of masks.

GAINING ATTENTION THROUGH EMPATHY AND DESIGN

Thus far in this chapter, you have looked at strategies for persuasion centered in logical and ethical appeals, both personal and community-based. But implementing these appeals often depends, first, on getting an audience's attention. That may be the hardest part, particularly when you're addressing a broad public audience. The third persuasive strategy identified by Aristotle is often what's needed. That appeal is *pathos*, an appeal to the audience's emotions and their aesthetic sense. Through empathy with them, you gain their attention.

Appealing to Your Audience's Emotions

Emotions are easily dismissed in most discussions of technical and scientific communication. But you gain an advantage, and serve your audience, when you design your messages with their aesthetic and emotional impact in mind. It's another way you adjust your ideas to the audience.

When scientists talk about the "elegance" of a theory, they describe a quality that is more than the mere accumulation of evidence and clever reasoning. The aesthetics of the argument may be as persuasive as its logic or the persona of the author. When several researchers simultaneously arrive at the same explanation of an occurrence, the one who reports it most effectively and attractively is often credited with the discovery. The lesson for you is clear. You need to think well and express that thinking in an appropriately clear and engaging style to make an impact.

Designing Digital Surfaces

Communication researcher Richard Lanham offers useful advice for developing that engaging style. It is based on a concept he calls "the economy of attention" (Lanham, 2006). In a world of too much information, too many messages, attention is the rare resource. To explain strategies that might help writers gain an audience's attention, Lanham first makes a distinction between two aspects of any digital text: *surface* and *content*. He argues that readers go back and forth (he calls it "toggling") constantly between the surface delivery of the message—that is, the screen display—and the content behind the surface. In his terms, readers toggle between looking *at* and looking *through*. For most technical and scientific communication, the surface aims at transparency. Style and delivery are often conventional, in the best sense. They make it easy to see through to the elegance of the argument itself.

But even when you communicate with colleagues, and especially when you communicate with a wider audience, a community less familiar with your work, you need to gain the audience's attention before you can persuade them. Given your understanding of surface and content, you have multiple options in today's multimodal environment for enriching both. Draw your audience into your message with a compelling digital surface, including images, motion, music, color, catchy phrases in interesting typography. Looking *at* these multimodal enhancements, an audience can be encouraged to look *through* them, in Lanham's terms, to the content.

Designing Content through Empathy

This chapter has discussed a variety of strategies for explaining and persuading. These incorporate both traditional genres and approaches and others that are coming into prominence, in part reflecting new technologies and community concerns. One, as you read earlier in the chapter, is using stories to explain technical content to non-expert audiences. Participatory community research, among other methods for developing content, aims at collecting stories of residents to foster beneficial changes in the community. Stories draw attention. They can also persuade audiences, who see others they trust talking about their own situations, to buy into a policy or behave in a way you advocate. The project on story mapping you read about in Chapter 7 centered on such narratives, along with maps, to carry its appeal. The message for residents of the low-lying coastal regions in the authors' study was simple. You can play a role in mitigating the effects of sea level rise in your region.

For more about community research, see Chapter 6

For more about the SLR project, see Chapter 9

As a technical and professional communicator, be aware of the wide range of strategies available to you as you develop content for the multiple audiences you address now and will address in your career. The next chapters in Part 3 provide more detailed advice for composing visuals and text to implement these strategies.

In this chapter you have learned basic strategies for explaining and persuading, two major purposes for communicating scientific and technical information. When you explain, you help your audience understand information or enable them to do something. When you persuade, you adjust your ideas to the audience and the audience to your ideas. Three strategies for persuasion identified by the Greek philosopher Aristotle are still productive: logical, ethical, and emotional appeals. In a *logical appeal*, you supply good reasons to convince your audience to accept an assertion you make—for example, when you argue that a particular cause produced a particular effect or might do so in the future. To win the argument, you also point out any errors in fact or reasoning made by an opponent. Your good character and instinct for community help you create common ground for persuading others through an *ethical appeal*. To draw audiences to that common ground, design an aesthetically pleasing digital display and develop content through empathy with your audience, strategies that demonstrate the *emotional appeal*.

CHECKLIST: EXPLAINING AND PERSUADING

1. Consider your purpose in communicating

    To explain

        Help the audience understand
        Help the audience do

    To persuade

        Encourage agreement with your idea or decision
        Encourage the audience to act in the way you recommend

2. To explain, help your audience understand by describing, classifying, and
defining
    Describe a mechanism, organism, site, or process
        Start with one or more visuals
        Add text to adjust the visual to your audience's prior knowledge
        Keep your perspective consistent

    Classify a group of items to foster audience understanding
        Identify common elements in abstract or natural phenomena
        Apply the principle to all items

    Define terms you expect the audience not to know or you use in a special
        sense
        Adjust the definition's length and location to your emphasis and your
            audience's prior knowledge
        Control the display of your attitude
        Use familiar terms in the definition

3. To persuade, adjust your ideas to the audience and the audience to your
ideas
    Provide good reasons in a logical appeal
        Think in terms of analogies
        Analyze cause and effect

    Uncover errors in the arguments presented by others
        Faulty generalizations
        Implied assumptions
        Either/or thinking

    Avoid errors in your own reasoning

4. Create a persona that demonstrates your good character and instinct for
community
    Reflect the personal qualities your audience values
    Show your knowledge of authorities the audience respects
    Acknowledge your point of view
    Understand culture differences in persuasion
    Create common ground by understanding others' values, norms, and
        perspectives
    Craft communications that demonstrate how each of you supplies what
        the other lacks

5. Appeal to your audience's emotions
    Design an aesthetically pleasing digital surface to attract audience
        attention
    Develop content that demonstrates empathy with your audience
    Engage your audience as an agent with you in advocating for positive
        social change

EXERCISES

1. For-profit and non-profit organizations, governments, and other institutions employ words, graphics, videos, and physical artifacts to persuade viewers and readers to adopt behaviors or accept ideas relevant to environmental sustainability practices. Here are some examples:

   - Internal corporate communications to employees concerning the organization's commitment to environmental sustainability
   - Corporate public relations communications to media, activists, regulators, or shareholders regarding sustainability
   - Taglines encouraging environmentally friendly practices (e.g., "Green Living Promises Longer Life," "Love Your Kids and Go Green")
   - Written reminders, posted on walls and hallways, to conserve water and electricity and to recycle packaging materials
   - Electronic messages at ATM machines requesting customers to refrain from printing transaction records as an environmentally friendly act
   - Wall art, including large plantings of flowers and grasses, intended to persuade viewers that corporations support sustainability initiatives
   - Labels and graphics on recycling bins.

   These can be ethically ambiguous. When, for example, a hotel requests guests to consider carefully the need for daily clean towels or linens, is the hotel management seeking to preserve the environment or to reduce laundering costs? At your instructor's direction, collect a few examples of public appeals to sustainability from your own digital and physical environment to share with the class. Be prepared to talk about their persuasive strategies (courtesy of Sam DeKay).

2. Analogies help you think about phenomena and explain concepts or theories, such as the Swiss cheese model mentioned in this chapter or an economic model called "donut economics." You'll find these often in newspapers, online magazines, blogs, podcasts, Ted talks, and other public-facing communications from scientific experts. Review your textbooks, too, for analogies that aim to instruct you in disciplinary practices and help you understand major theories. Then, write a brief report about the role of analogies in science, technology, and other disciplines you are studying.

3. What assumptions are implied in the following statements?

   - People who like their work ought not to get good salaries.
   - *Environment* is a broad term that embraces problems of air and water pollution, birth control, urban transportation, and everything that affects or threatens human survival.
   - We can no longer afford the luxury of designing for conditions that do not exist.

FOR COLLABORATION

Depending on the size of your class, divide into teams of three or four. Each team will strategize for a part of the class period on the problem of encouraging students toward positive and productive engagement with social media. Such engagement includes, for example, overcoming confirmation bias, uncovering misinformation and disinformation, knowing what you're signing up for when you sign onto various platforms, being aware of surveillance strategies—the list goes on. How do discussants or influencers on social media establish their credibility? What values are taken for granted in what you (or an algorithm) select to read online? What kinds of proof are offered for assertions?

On your team, explore your own experiences in the argumentative culture of social media. Then, devise a communication strategy to persuade fellow students toward effective use. Pick one chief problem to address You need to be brief and probably use high visual impact, maybe in an infographic (see Chapter 9). Sketch it out at this point. Then, at the instructor's direction, each team will share its sketch and its strategy in a short talk with the class.

REFERENCES

Christakis, N. (2020). The Swiss cheese model for combating Covid-19. *Wall Street Journal.* November 14–15, C1.

Grant, R., & Fisher, K. (1971). Scientists and science writers: Concerns and proposed solutions. *Federation Proceedings* (Federation of American Societies for Experimental Biology), *30*, 819.

Kaplan, R., & Frosch, D. (2020). Honesty is the best policy in selling vaccines. *Wall Street Journal.* December 29, A15.

Lanham, R. (2006). *The economics of attention: Style and substance in the age of information.* Chicago, IL: University of Chicago Press.

Page, A.N. (2003). Review of Nisbett, R.E. *The geography of thought: How Asians and Westerners think differently and why. Business Communication Quarterly, 66*(3), 138–142.

Tingley, K. (2020). Could better public health messaging persuade more people to change their behavior in ways that would curb the pandemic? *New York Times Magazine.* 13 December, 14–16.

Voltin, S. (2020). Christopher Newell. At the summit. *Portland Magazine.* October, 22.

Young, R.E., Becker, A.L., & Pike, K.L. (1970). *Rhetoric: Discovery and change.* New York, NY: Harcourt, Brace, & World.

9 Composing Visuals

As a technical professional, you'll communicate as much in visuals as in words, sometimes even more so, depending on your audience and purpose. Especially with empirical research, you collect and generate information largely in numbers and images. Visual devices help you observe well and document what you see. Drawings and sketches also foster a team's iteration on the problems, solutions, prototypes, and opportunities addressed by your project. Many great buildings began as a few lines drawn on a napkin. Today's multimodal environment for communication offers abundant options for you to record your thinking and communicate with audiences through visuals and design.

For more about empirical research, see Chapter 6

This chapter, in a monochrome print publication, cannot reproduce the full-color digital environment you spend time in as a student and will in your career. It can, however, alert you to strategies for incorporating the affordances of a multimodal approach into your communication products. Visuals enhance your ability to explain something and to persuade audiences to accept your evidence and ideas, even act on them. This chapter continues the discussion in Chapter 8 addressing these broad communication purposes. While you are reading, keep an eye out for examples and models of good practices described in this chapter on the screens you see every day. What you learn in the sections that follow should also help you better observe the impact of images and design as you experience the natural and the built environment at large. Those observations, along with the design-thinking-enhanced strategies for communication discussed in this chapter, will stand you in good stead as you compose effective and attractive visual content for your audience.

THE BASIC FORMS

Figure 9.1 provides an overview of visual forms commonly used to explain and persuade in technical and professional settings. You're probably familiar with all of these forms. They range from the relatively concrete (such as a photograph and some drawings and maps) to the more abstract, such as an organizational chart or a schematic diagram. Later in the chapter, you'll read more about the abstract end of the range, including graphic design. These forms can also be mixed and matched in ways that both clarify and confuse. Digital technology has greatly enhanced your ability to create visuals and graphics. But being able to create some effect should not, alone, be the deciding factor in determining what to produce for an audience.

DOI: 10.4324/9781003093763-12

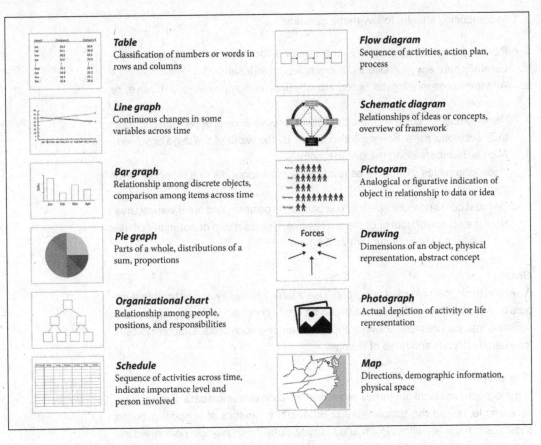

Table
Classification of numbers or words in rows and columns

Line graph
Continuous changes in some variables across time

Bar graph
Relationship among discrete objects, comparison among items across time

Pie graph
Parts of a whole, distributions of a sum, proportions

Organizational chart
Relationship among people, positions, and responsibilities

Schedule
Sequence of activities across time, indicate importance level and person involved

Flow diagram
Sequence of activities, action plan, process

Schematic diagram
Relationships of ideas or concepts, overview of framework

Pictogram
Analogical or figurative indication of object in relationship to data or idea

Drawing
Dimensions of an object, physical representation, abstract concept

Photograph
Actual depiction of activity or life representation

Map
Directions, demographic information, physical space

9.1
Basic Forms
for Presenting
Technical and
Professional
Information

Select visuals and graphic design options critically to represent your own or your organization's persona and your understanding of the audience's needs.

Table

One simple, if powerful, form is a table. In a table, you arrange words or numbers in rows (horizontal) and columns (vertical). The periodic table of the elements is a classic example, reproduced on mugs and t-shirts in its original form or as a meme incorporating foods, flowers, or many other elements. Spreadsheets are a common management and financial application of a table. Tables serve several purposes:

- Recording numerical information consistently, as in a price list
- Identifying holes in your information where cells (the intersections of rows and columns) are empty
- Showing many discrete units of information in a small space
- Making it easy for readers to compare items
- Presenting a series of parallel terms as a multidimensional list.

Most tables read vertically—that is, in the same direction as the text (in "portrait" position). Long tables, however, may run horizontally ("landscape" position). To show a network of interrelationships among items, either words or numbers, use a matrix, another form of table.

In composing a table, follow these guidelines:

1. Place units of measurement in the title or a footnote, not after each entry. Carefully note any multipliers (for example, "in thousands").
2. Arrange items in a logical order: alphabetical, geographical, quantitative, or chronological.
3. Place numbers in columns rather than rows for ease of comparison and carryover to an additional page. Rows are also limited by the width of the page or screen.
4. Align all numbers along the decimal point.
5. Place long tables that serve as references in an appendix to a report or other document.
6. On most occasions, use spacing to separate the columns and avoid vertical lines.
7. Note the source of data or of the entire table. Use a system of notation consistent with that of your document as a whole.

Graph

A second basic form of visual is a graph. Use a table to classify and assemble numbers or words so those discrete units can be easily compared. Use a graph to turn numbers into pictures that show relationships among such units. Graphs indicate, for example, trends and rates of change.

Line Graph

In a line graph, you highlight trends and relationships in numerical data. Such graphs, for example, record the temperature or other characteristics of hospital patients, show customer preferences in a market survey, detail how a vehicle performed in a series of engineering tests, or summarize a company's financial picture. Computers that monitor experimental data often indicate results directly in line graphs. Some numerical values lend themselves to expression in both a table and a graph. For an audience who needs the details, use a table. To show a trend to an audience who needs an overview or needs to be persuaded, use a graph.

In composing a line graph, follow these guidelines:

1. Show the independent variable (often a measure of time) along the horizontal axis (the x-axis).
2. Show the dependent variable on the vertical axis (y-axis).
3. Choose a scale that well represents the importance of the changes shown.
4. For ease of reading, use multiples of 2, 5, 10, and so on.
5. If you need to, include an insert that shows an important segment at magnified scale to supplement a line that covers a longer time. For example, a plot of how a specimen's weight changed over a 2-day period might include an insert showing significant changes over the first few hours.
6. Use a series of graphs with parallel scales if you need to show more than two or three lines.
7. Present information as it was originally recorded. If the recording was continuous, don't show individual data points.
8. Keep each line equal in width, because one that is wider than the others appears to be more important.

9. Label the vertical axis to the left to keep the area within the grid as free of text as possible.
10. Limit explanatory material on the graph. Save descriptive discussions, such as test conditions, interpretations, or apparatus, for the text.

Bar Graph

Use a line graph to show a trend or a continuous progression over time; use a bar graph (sometimes called a bar chart) to compare the relationship among discrete items. For example, connect each month's statistics about births in a community in a line that shows at a glance fluctuation in the birth rate over a year. To emphasize the comparison from month to month, use a bar graph. The length of the bar indicates a value (for example, 80 percent) or amount (for example, 10). You can arrange bars either vertically or horizontally. Vertical bars often represent changes over time, with the baseline acting like the horizontal axis of a line graph. Horizontal bars commonly show acceleration or percentages.

Bar graphs appear frequently in corporate annual reports, general interest magazines and newspapers, and other publications aimed at a broad readership, because these graphs are easy to understand. They can be misleading, however. You can create attention-getting graphs that cause the reader to come to the wrong interpretation.

In composing a bar graph, follow these guidelines:

1. Use solid bars or differentiate bars with various colors, hatching, or perspective.
2. Stack information within one bar to compare data from different years in one category, or cluster the bars.
3. Include bars at full length to avoid distortion (unless the longest bar is really irrelevant to the interpretation).
4. Begin the scale at "0" so the height accurately represents the value.
5. Avoid a three-dimensional look unless you intend to indicate that both the length and the volume of the bar are meaningful. Otherwise, the reader may compare volume instead of length.
6. Similarly, maintain equal bar width and spacing so the reader doesn't think the width is as informative as the length.

Pie Graph

Use a pie graph, sometimes called a pie chart, as the name implies, to divide up a total (the pie) into pieces. That total may be, for example, a sum of money or a sample population—all students at your university, all civil engineering majors, all residents of a particular city. Pie graphs are common in articles and reports for nonspecialist audiences but less common in technical discussions. They often take up more room on a page or screen than the information deserves, so use them sparingly.

In composing a pie graph, follow these guidelines:

1. Begin the division of the pie at the top center ("noon").
2. Arrange sectors by size, with the largest at the top.
3. Place labels either inside or outside the circle—preferably without lines that tie the label to the pie.
4. Label slices with percentages, not just words.
5. Try to show no more than six segments.

Diagram

A third basic visual form is a *diagram*, sometimes called a "chart." A diagram may represent numerical information, like a graph, but it can also show other relationships, from the relatively concrete to the relatively abstract.

A diagram may be both descriptive, showing what happens, and prescriptive, showing what ought to happen. Prescriptive diagrams accompany instructions for assembling devices, operating equipment or systems, scheduling a project, or reinventing a management structure. In the process of drawing a diagram on a whiteboard or other vertical surface, teams ideate on projects to plot their way collaboratively to effective decisions about a course of action.

For more about ideating with diagrams, see Chapters 4 and 5

Organizational Chart

An organizational chart shows how people or parts relate in a structure. The most common form is the organizational chart of a corporation or agency.

Schedule

A schedule helps individuals and teams plan and monitor activities. Your class schedule, for example, both describes when you'll be in class and prescribes your activity for the academic quarter or semester. Contractors use schedules to ensure that the right people and equipment arrive at the job site at the right time. Researchers include a schedule of planned project activities in a proposal seeking funding to support those activities. The most common form of schedule is probably the Gantt chart: tasks are listed in the rows. They are measured against time units designated in the columns.

For more about schedules in a proposal, see Chapter 13

Workflow schedules are often complex, engaging a large number of people to be coordinated in a multistep process in uncertain and changing situations. For that reason, developers have created a variety of software packages to create such schedules.

Flow Diagram

A flow diagram, as the name implies, shows a flow or process. The diagram may use text and drawings to describe something concrete. Or it may represent abstract thinking, as in a decision tree. In such diagrams, connect symbols with arrows to indicate the direction of flow. Such arrows are especially important internationally, because people who read from left to right (as in English) tend to see a sequence as beginning at the left and moving right. People who read right to left depict flow in that direction too.

Schematic Diagram

A "schematic" demonstrates electrical or mechanical connections in a series, as in a circuit or computerized system. An overview schematic may be supplemented by more detailed diagrams for manufacturers or field crews.

Pictogram

A pictogram enhances the pictorial possibilities in any diagram or graph to motivate or entertain a popular audience. The bars of a bar graph, for example, may suggest the topic under discussion: people, bags of money, hockey sticks, or computers.

To persuade patients to complete the course of an antibiotic even when their symptoms disappeared, for example, a British physician drew a chart that showed spoons running to a finish line across hurdles representing each dose. The spoon that crossed the finish was the winner. Several patients, however, objected to seeing their health practices as competitive. Such misreadings are a potential problem with pictograms. So is bias or regionalism in the symbols—for example, in the shape of electric plugs, which differ globally.

Drawing

Drawing pictures is an age-old way to show an object or place. The process now is commonly a function of software that composes images from data that you input, a process that is easier and more precise than freehand composition. The output is often called a "rendering." In a drawing, you can show more dimensions of an object or place than is possible in, say, a photograph. You can show both the surface and what's beneath. A photograph would not let you see inside in the same way.

Readers of drawings need to understand the conventions that underlie them. One convention is representing the whole by a part. When you show only the head to represent a whole animal, for example, be aware that some readers may not understand that convention. Drawings can look realistic, although some people have trouble finding an image in a series of discrete lines.

Photograph

A photograph shows how something actually appears, in its context. Many readers find a photograph easier to understand than a drawing. Through a photograph, for example, a researcher can examine a laboratory setup that she may want to duplicate. Photographs of authors often accompany articles in journals and magazines. They help readers visualize who is speaking to them in the article and remind them of colleagues whose faces they may recognize but whose names may be forgotten. A computer-generated image can also show things that are too big, too small, or too remote to be seen by the unaided observer. With such a photograph, you can help your reader see into cosmic space, into the brain, or into the microstructure of an atom.

A photograph is often dramatic and draws attention, attractive characteristics when you want to persuade a reader. But a photograph has disadvantages. A photograph may show too much detail when the main object dissolves in a field of other objects. Photographers need to zoom in on their subjects, whether people or equipment. Equipment looks best when it is at work.

Details in the setting may raise unanticipated issues in communications that cross cultural or national borders. A brochure for a German conference hotel, for example, contained a picture of its swimming pool. In the background were two nude women, a detail that might offend more conservative Americans.

Map

The final item in this brief survey of basic forms of visual presentation is a map. In traditional usage, a map depicts landforms. A *chart* depicts water or sky. But that distinction is becoming blurred, except for mariners.

Attributes

All maps display three attributes: scale, projection, and graphic symbols.

- *Scale* shows how much smaller the map is than the reality it represents. Paper maps adhere to fixed scales indicated in words or numbers on the map itself. With computer-generated maps, users can vary the scale to zoom in on a point of interest in a dynamic flight through the landscape.
- *Projection* refers to the way a three-dimensional territory is represented on a flat plane. Projection techniques always distort, as in the Mercator projection world map that creates an overly large Greenland. But that map system worked well for Europe and North America in Mercator's time.
- *Graphic symbols* represent features on the map. Three types are common: point, line, and area. Point symbols, for example, note the location of cities. The relative size of a dot or circle by the city's name can also indicate the relative size of the city being designated. Pictorial points can add even more information as in a stylized skier representing a ski area. Line symbols show, for example, roads and rivers.

GRAPHIC DESIGN

Design thinking, as you've been learning about the process in this text, is a broad approach to solving problems and pursuing opportunities. It fosters better interactions between humans and their environment in a wide range of situations. It extends a discipline practiced for years by artists, architects, craftspeople, and others concerned with planning and prototyping products and processes for such settings as homes, landscapes, offices, and industry. To design, as one cliché goes, is to bring order out of chaos. If you are a student in engineering or science, you may be collaborating on projects with students in an art and design or visual communication program. Such collaborations are becoming more common in academic makerspaces and studios, as well as in professional communities.

In addition to using the common vocabulary of forms for presenting information in visuals reviewed first in this chapter, you should also become familiar, if you aren't already, with graphic design more broadly. In addition to photography and illustrations, this includes typography, color, and iconography. Your choice in these elements of graphic design contributes to the persona you establish in your communication and reinforces your personal brand. When you write as a representative of an organization, you adhere to standards for its brand, often in a design program developed by an advertising or marketing agency. The program usually begins with the development of an organization's logo, the stylized trademark or signature that represents their name, such as the Nike swoosh. It then incorporates consistent choices within the other three elements described in this section.

Typography

A major element, both in print and especially online, is typography—that is, typeface and font. Dedicated software allows designers to create an almost unlimited range of type, each with its own personality, creating its own aura. The standards,

Times New Roman and Calibri, still work. But a quick scroll on your computer's menu of font collections will probably provide more options than you'd like to face, each with fonts in bold, italics, and underlining.

Times New Roman has *serifs*, little extenders (like feet) on the ends of the vertical and horizontal strokes of letters. US readers have traditionally found such type easy to read. In Europe, however, *sans serif* (without serifs) has been more common. But that's changing. Calibri, often the default on Apple products, is a sans serif type. Released to the public in 2002, Gotham, a geometric sans serif type, has become widely popular online, on billboards, on products, and on buildings. According to its designer, it has a "straight forward voice—even stark" (quoted in Hawley, 2019). Its letters incorporate basic shapes that make it easy to read at a distance. It advances a trend toward minimalist design useful in multiple applications worldwide.

If you now take the look of type for granted, look again. Use typography to clarify information, perhaps to have an emotional appeal, and to establish your brand.

Color

Like typography, color plays an important role in visual communication. Computer-generated maps of the seafloor, for example, use color to indicate depths and elevations, from deep blue for the greatest depth to red for the higher elevations. Shading adds an almost three-dimensional element. Scientists discussing the structure of cells often use color to code different filaments of interest. Color helps identify individual lines in a graph and key them to a legend that provides more explanation about the source and meaning of the line. It adds a dramatic element to tables and matrixes, which otherwise might seem dull.

Color can also be instructive. For example, electronics manufacturers often color-code the cables and sockets on their machinery. The colors ease the job of instructing customers in setting up the equipment. Rather than having to describe cable locations, the manual simply notes, "Plug the purple cable into the purple socket." Rustproofing paint applied to metals that must endure harsh weather, such as deep-sea oil rigs, bears different pigments for different layers of application. Maintenance staff can discern easily how much wear the surface has endured by the color of the exposed paint layer. The buoys that identify ownership of lobster traps in Maine are color-coded. The colors are registered on the lobstermen's licenses. In the water, lobstermen can easily identify their own and others' traps.

Color is dramatic and grabs attention. But its impact varies across cultures and professional disciplines. Red, for example, is associated with danger in North America, but with joy and festivity in China. Americans wear black at a funeral, connoting mourning, and white at a wedding, connoting hope. Many in India wear white at a funeral, connoting purity and the eternal status of the soul. In New Mexico, windowsills are often painted blue because the color seems to repel mosquitoes that would otherwise enter the house. Similarly, the Japanese traditionally see blue as connoting villainy, but to Americans, blue often seems positive, calm, and authoritative. One researcher notes that different professional groups interpret colors differently, as shown in Table 9.1.

Table 9.1
How Different Professionals Interpret Colors

Color	Financial Managers	Health Care Professionals	Control Engineers
Blue	Corporate, reliable	Dead	Cold, water
Green	Profitable	Infected, bilious	Nominal, safe
Red	Unprofitable	Healthy	Danger

Because color may miscue readers who do not share your culture or profession, test color choices with a sample of users before you finalize a communication product.

Color introduces other problems as well. Some 11 percent of men are color-blind, and so color as a device for differentiating items in a text is meaningless for them. In addition, although most people can discriminate a wide range of colors, if their culture doesn't support the recording of that discrimination with terms for each shade, the distinctions may go unnoticed. For example, the Japanese make subtle color distinctions unrecognized in other communities. These distinctions, called *shibui*, depend on a combination of color and texture. Be aware, too, that you may prepare your text on a color screen, but it may see final form in a monochrome print publication, like this book. Make sure that core meanings are clear without reference to colors. Color is a powerful design tool, to be used with some caution.

Iconography and Symbol Signs

Traditionally, an icon is a venerated religious image, but the term has taken on a different meaning in graphic design. On a computer screen, an icon represents a function to be performed: a clipboard for "paste," a printer for "print," a waste basket for "move to trash," the handset of a phone for "make phone call." Figure 9.1 represents the basic visual forms as small icons. Graphic symbols also serve to describe processes such as package handling, laundering, household maintenance, or the assembly of items such as toys and furniture. On highway signs, they designate the locations of such facilities as hospitals, rest rooms, swimming pools, and eating places. Road signs, including stop signs and pedestrian crossing marks on pavements, govern the behavior of vehicles and pedestrians. The designer you met in Chapter 1 proposed to help elderly passengers find their way through the airport by enhancing the signage. Such environmental graphics are responding to an increased awareness of the need to serve diverse communities regionally and globally. They help speak an international language.

PERSUADING WITH IMAGES

The six basic forms—table, graph, diagram, drawing, photograph, and map—help you turn the information you collected in your research into visual content that explains something or provides supporting evidence for an idea. As you saw in Chapter 8, today's multimodal digital environment also offers new structures and genres for gaining attention and persuading audiences. They help audiences visualize or otherwise experience phenomena, interactively, sometimes in the three dimensions of augmented reality (AR). An international furnishings store, for example, offers an AR app to its customers so they can preview how new purchases would look in their own home. Often derived from, or riffing on, these basic forms, especially photographs, videos, illustrations, and maps, these new visual genres offer strategies for engaging audiences with images that tell stories and present scenarios. In implementing these strategies, you can evoke a personal brand or persona as someone who is well informed, empathetic, and worthy of trust. Here are some examples of strategies in action. Be alert to other models you can imitate or adapt as you encounter screen displays and environmental graphics every day.

Infographics

An infographic is a poster-like image with content divided into a layout of sections. The aim is to help an audience visualize complex data in one super image available at a glance (Figure 9.2).

Publishers of professional journals are recognizing the important role infographics can play in helping their authors influence government agencies, policy makers, industry professionals, the public, and the media. Sage, for example, publishes journals mainly in the sciences and social sciences whose articles present highly technical research for audiences of colleagues in the authors' field. To impact a broader audience, the publisher's staff offers its services to authors to convert the research content and core findings into a single image, which is "visually appealing and shareable." It aims to better compete with what Sage notes is the "4000 branded messages" circulating daily online.

In addition, many technical and scientific conferences include poster sessions as part of their program. Researchers, both students and professionals, stand next to a large-format poster on an easel that presents their work. They discuss its implications with conference attendees walking past. Infographics are also common on social media, particularly Pinterest, Facebook, Instagram, and Twitter. Templates for creating them are widely available commercially online.

For more about creating posters, see Chapter 19

9.2
An Infographic Summarizing a Published Study

Source: Courtesy of Sage Publishing

In general, keep these guidelines in mind as you create an infographic:

1. Keep it simple. The tendency toward what one graphics expert calls *chartjunk* is high (Tufte, 1990).
2. Use only bite-size units of text.
3. Allow adequate white space.
4. Use a large-enough font for easy reading.
5. Keep font choice consistent.
6. Have one central message and subordinate information to that message.

Videos

Along with other publishers, Sage also offers its services in composing video abstracts of articles. These include animation, voice-over commentary, background music, and short bits of text, aimed to extend the author's research into the public and policy-making domains. In Chapter 6, you read about the use of cell phone videos in participatory community research. Those videos, in turn, can become part of a communication product, as you'll see in the next section.

Story Maps

Throughout history, maps have been important devices for communicating information about place and about events and people in places. Increasingly, statistical data have also been productively visualized through computer-generated forms based on a mapping concept. Data visualization is a fast-growing scientific field. Advances in technology have also led to the creation of interactive maps that layer extensive information about land forms and the sea and adjust dynamically to a user's requests. These digital maps, however, can be highly detailed, require specialized skills for their interpretation, and may overload non-expert users.

A new map-based approach to presenting environmental information in a more user-friendly and persuasive way is the story map. In Chapter 7, you read segments from a research article describing a project to develop a story map. The authors

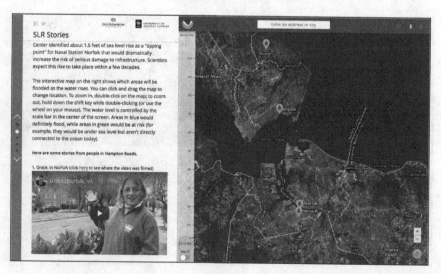

9.3
Screenshot of a
Story Map

Source: Stephens and
Richards (2020)

created a tool aimed to alert residents of low-lying coastal areas about the risks of sea level rise. In a community-based approach, they interviewed residents in two areas on the east coast of the US who participated with them in the project. They then used a template commercially available online to combine standard as well as interactive maps with videos in which participants told stories of their lived experience with flooding. The template, a form of *middleware*, is relatively easy for both researchers and the public to interact with. Figure 9. 3 reproduces a screenshot from the researchers' interactive tool (you can access the project by visiting https:// tinyurl.com/slr-stories).

The content of the Stephens and Richards story map includes:

- Maps of the two regions, some of which are interactive
- Explanatory text about how to use the tool and about the project, including a concluding segment urging action
- Several 3-minute videos (culled from 10–25-minute interviews, filmed outdoors, on-site) of residents telling their stories about personal encounters with sea level rise.

Environmental Graphics

Story maps depict environmental concerns in multimodal ways on a screen. In another approach, experts are collaborating with artists to create murals and other open-air presentations aimed at encouraging popular understanding of scientific phenomena. An example is the initiative at the Wisconsin Institute for Discovery called Science to Street Art, which "pairs scientists and street artists to collaborate with communities to create science civic art" (Contreras, 2021). One mural produced through this initiative consists of a Wisconsin landscape scene on which magnified images of unseen molecular structures are superimposed (Figure 9.4).

When part of a building was torn down in Newark, DE, to create public space around a new bookstore, faculty at the University of Delaware proposed that the

9.4
Mural on a
Sewerage District's
Pumping Station in
Madison, Wisconsin

Source: Molecular
Structure Mural—
Science to Street Art,
Wisconsin Institute for
Discovery & Madison
Metropolitan Sewerage
District collaboration
Artists: Ingrid Kallick,
Peter Krsko; Scientists:
Jo Handelsman,
Zachary K. Wickens;
Photographer: Morgan
Eder

Newark Public Art Map

This interactive map enables viewers and community members to take pride in and virtually explore public art within the City of Newark (Del.) and art displayed on the University of Delaware's campus.

Delaware Complete Communities Toolbox

| All Newark Art | City of Newark Public Art | UD Campus Art |

12 The Wall

Artist Name: Diamond State Masonry: William Austin Jr., Christopher Brown, Augustine Cerminara, Keith Gainey, Gerald Giammatteo, George Hall, Thomas Kellems, Paul McConomy, Brian Pazdalski, James Satterfield, Robert Snyder, and Joseph Woolman

Year: 2011

Artwork Type: 3-D

New Castle County, Delaware FirstMap, Esri, HERE, Garmin, INCR...

blank wall left in place be repurposed as a "learning wall." The campus architect and others, including a geographer specializing in historic houses, designed the surface to illustrate ten different bonds for aligning bricks. A team of highly skilled masons completed the design. A photograph of the wall is included in an interactive story map about public art in the town. The map explains the bonds and the concept of the wall (Figure 9.5).

9.5
A "Learning Wall" in Newark, Delaware
Institute for Public Information, University of Delaware

COMPOSING EFFECTIVE LABELS

You don't need to label visuals used primarily to enhance the atmosphere or aesthetics of a communication product. You should ensure that visuals are as self-explanatory as possible. But conventions dictate that you label any visual within a more traditional genre. The approach and extent of the label depend on your purpose, the audience's familiarity with the form of the visual and its content, and the genre of the presentation. In relatively informal communication, a short title may identify the people or objects in a picture or add a humorous twist. In a technical or scientific article or report, however, the visuals may be the main content, and labels provide commentary that explains and connects them. Such labels usually consist of three parts: a number that shows the sequence of that visual in the report, a brief title, and an explanation.

The number—for example, *Figure 1*—helps you and the reader refer to the visual in a long document. The label helps you transform the data you collected in your research into evidence that enlightens or persuades someone else. Delete any internal codes or identifiers that are meaningless to the reader—for example, "Run 11," "Alloy 60," or the date of the run. The length and content of the explanation depend on the audience's sophistication and the complexity of the evidence. If the visual was originally created by someone else or uses data someone else compiled, your label should also include a reference to that source:

> . Source: Science Resources Studies Division, National Science Foundation, Research and Development in Industry: 2020 (Washington, DC: NSF, 2020).

Or you may place a reference notation in the label consistent with notations used elsewhere in the report:

> Figure 3. Lobster Landings in Casco Bay, Maine 2020–21 [6]

In a report or article, introduce every visual in the text. None should appear unannounced. The least invasive mention is a parenthetical one:

> The beginning of each heart beat is detected as a large, positive spike in the differentiated pressure waveform (Figure 2).

To draw even more reader attention, use a full sentence that instructs the reader. Avoid simply pointing to the visual:

> *Poor.* Figure 2 shows a typical waveform.

Instead, highlight it:

> *Better.* As shown in Figure 2, the beginning of each heartbeat is detected as a large, positive spike in the differentiated pressure waveform.

COMMUNICATING ETHICALLY WITH VISUALS

When you represent information in a graphic form—interpreting, condensing, illustrating trends and relationships—you also distort the reality of the data. That's the point. You are turning data—raw results—into information—results you've interpreted. Finally, you turn those interpreted results into visual content to present to the audience. The value of the visual, however, is how well it generalizes and enhances the audience's understanding. You learned in Chapter 8 about how readers of presentations on screen toggle between the surface and the content. Technical and scientific presentations aim at transparency, no less with visuals than with text. The design of your visual should be true to the data it presents.

The clearest form, in this sense, is a table for presenting numerical data. Although you classify and line up the numbers, the numbers themselves are still there for an audience to interpret differently if they like. In graphs, you aim to show relationships. In doing so, either by accident or deliberately, you can design a graphic that uses a misleading scale, distracting pictures, an excess of lines or colors, superficial complexity in design features, or other devices that ultimately confuse the audience and hide or distort the truth of the data. This is *chartjunk*, in the term of a leading authority on visuals (Tufte, 1990). The surface presentation misleads the audience about what the data mean rather than enhancing their understanding. The result is: You lie.

Computer-generated photographs in particular can cross ethical borders when they are manipulated digitally to provide false evidence. Viewers tend to believe that photographs show something real. But you can now, relatively easily, create images of composite faces, eliminate unsightly features from a landscape, show a building where none exists, and otherwise depict a reality that may mislead. It's also easy to share manipulated images online, in social media.

For more about manipulated images, see Chapter 3

Be careful, too, with maps. As you know, geographic information system (GIS) apps are richly flexible ways to collect and represent all kinds of information about people and how they move through places. The apps also help you find your way through places. Today, researchers are learning that young people are losing their innate ability to located themselves in a neighborhood or city because they depend instead on such tools. Many of us would be literally lost without GIS devices on boats and cars and smartphones. But maps are also political tools, sometimes with unethical consequences. Government leaders sometimes use maps as power plays to claim territory or to show the riches of the territory they already possess. Military agents may use maps to misinform their enemies. A story circulated in 2019, for example, that, on Google apps downloaded by Russian users, the then-disputed territory of Ukraine, held by Russians, was shown as part of Russia. Downloaded by Ukrainian users, the map showed it as part of Ukraine. Whether the story is true or not, the possibility that a platform could be manipulated in this way is very real (Higgins, 2019). As with all your communication, in visuals and in text, maintain your instinct for the ethical treatment of your audience.

In today's multimodal environment for communication, an ever-expanding range of strategies for presenting information visually is available not only to explain something, but also to persuade audiences to accept your evidence and ideas, even act on them. This chapter has briefly reviewed some common forms and strategies. Avoid using them simply to dazzle or make bad information look good. Use them ethically, effectively, and creatively to enhance your audience's ability to understand and to act.

CHECKLIST: COMPOSING VISUALS

1. Use visuals to explain and persuade
2. Select the appropriate visual form to explain information to your audience
   Use tables to compare discrete units, in numbers or words
   Use graphs to turn numbers into pictures that show trends and relationships
   Use diagrams to show relationships, from the concrete to the abstract
   Use drawings to show multiple dimensions of an object or process
   Use a photograph to show how something actually appears, in its context
   Use a map to show landforms, or, increasingly, to visualize stories and data

3. Use graphic design to reinforce your persona and clarify your content
      Select the appropriate typography to represent your message
      Use color to draw attention as well as instruct and enhance content
      Include icons and other signs to signal directions and speak internationally

4. Use images to persuade
      Use an infographic to help an audience visualize complex data at a glance
      Use multimodal strategies, including videos and music, to gain attention
      Use story maps to enlist a community in advocating for positive change
      Use environmental graphics to encourage popular understanding of science

5. Use visuals ethically to enhance an audience's ability to understand and to act

---

EXERCISES

1. At your instructor's request, write a memo that responds to one or more of the following tasks:

    *   Examine a sample of the visuals in *The Wall Street Journal*, *Scientific American*, and a technical journal in your field. How do they differ? How are they alike?
    *   Examine the visuals in five corporate annual reports. They may be 5 years of reports from the same company or the reports from five different companies. Do the visuals aim primarily to convey information or to serve as decoration? What do they say about the company's character, its culture, or the reliability of its reports? How are visuals integrated with the text?
    *   Compare a photograph taken by a drone flying overhead, a topographic map, and a GIS display of the same geographic area. How do the three images differ? How might each contribute to a report?
    *   Find two articles on broadly the same subject, one in a technical journal and the other addressed to a broad public. What visuals does each author use to appeal to the audience? How much background in mathematics or graphic interpretation does the audience need to understand the visuals?

2. Explain the rules of some game—for example, soccer or baseball—to an audience unfamiliar with the game. Use only visuals.

3. Arrange into a table the following information about a proposed course on the feeding habits of wildlife: The lecture schedule is based on the 10-week quarter. Lecture will be 3 hours a week. The material is divided according to species. The first 2 weeks will concentrate on deer. Weeks 3 and 4 will be devoted to rabbits. The next 2 weeks will cover the three squirrel species found in the northeastern United States. One week each will be assigned to the pheasant, quail, and ruffed grouse. The last week of lectures will cover the wild turkey. Diets of adults, immatures, and reproducing females will be studied for all species.

4. The following statistics represent the sales per hour of pizzas at three stores in a college town on a Friday night:

    - 8–9 pm: X Pizza, 36; Y Pizza, 47; Z Pizza, 12
    - 9–10 pm: X Pizza, 28; Y Pizza, 54; Z Pizza, 56
    - 10–11 pm: X Pizza, 10; Y Pizza, 62; Z Pizza, 61
    - 11–midnight: X Pizza, 10; Y Pizza, 58; Z Pizza, 12.

    Illustrate these data in a line graph, a bar graph, and a table. Which version seems most effective to compare the effects of Z Pizza's late-night buy-one-get-one-free coupon? To compare sales at each shop? To show how sales change hourly?

FOR COLLABORATION

Continue or begin the collaborative exercise described at the end of Chapter 8. Create the infographic, poster, or other communication product aimed at persuading students to use social media effectively and ethically. High visual content is essential, but you may use some words.

REFERENCES

Contreras, G.A. (2021). Thoughts from the science to street art director. Retrieved from https://bit.ly/3kjnGQR

Hawley, R. (2019). Gotham. How this one font took over the world. *The Outline.com*. Retrieved from https://bit.ly/3u3Vaot

Higgins, A. (2019). Apple, bowing to Russian pressure, redraws its map of Crimea. *The New York Times*. November 29, A4.

Institute for Public Administration, University of Delaware. (2020) Newark Public Art Map. Retrieved from https://bit.ly/3dSkoQZ

Sage Author Services. (2021). Retrieved from https://bit.ly/2Pknkw2

Stephens, S., & Richards, D. (2020). Story mapping and sea level rise: Listening to global risks at street level. *Communication Design Quarterly*, *8*(1), 5–18.

Tufte, E. (1990). *Envisioning information*. Cheshire, CT: Graphics Press.

10 Composing and Structuring Text

You'll use visuals, as you saw in Chapters 8 and 9, to think about your project and convey your findings to others. Sometimes, visuals alone carry the message. But to explain and persuade, you'll also need to compose text. This chapter discusses how to compose effective units of text and structure those units in a communication product. It first looks into the basic units: paragraphs, sentences, lists, and headings. The chapter then addresses the larger structure into which you'll assemble those units to deliver the right content to your audience.

PARAGRAPHS

The paragraph is the most basic and important unit of text (we don't speak in paragraphs). As you group ideas in paragraphs, you clarify your own thinking and design units of content to match the capacity of the audience to understand and to pay attention. To work, a paragraph has to do two apparently contradictory things at once. It must stand still and it must move. It stands still in that it is unified and adequately treats the essential aspects of one point. It moves in that it begins at one point and develops toward another. Within each paragraph:

- Motivate the reader.
- Lead the reader from familiar to new information.
- Show hierarchy or parallelism.
- Use topic sentences and transitional devices.

Motivating the Reader

Composing your text in units of paragraphs motivates the reader to move through the discussion. Remember to create paragraphs, a guideline often forgotten by writers of email messages. Indenting the first word of a paragraph or double-spacing between single-spaced blocks makes an attractive page or screen. In addition, adjust the length of paragraphs to reflect stages in the development of your discussion. For example, you can place material that's easy to understand in longer paragraphs. But use shorter ones for more difficult content. Adjust the length, too, to allocate your reader's attention. Fend off monotony by avoiding a succession of short, choppy paragraphs or a dismal collection of long ones. Avoid especially using long paragraphs at the beginning and end of a discussion, segments that require special, attractive strategies.

DOI: 10.4324/9781003093763-13

10.1
Original and
Revised Paragraph

Original. Soils represent major sinks for metals like cadmium that are released in to the environment. Soil does not have an infinite capacity to absorb metal contaminants and when this capacity is exhausted, environmental consequences are incurred. Contamination of soils by cadmium and other heavy metals has become a global concern in recent years because of the increasing demands of society for food production, waste disposal, and a healthier environment. The main causes of cadmium contamination in soils are amendment materials (e.g., municipal waste sludge) and fallout from nonferrous metal production and power plants.

Revised. Such sources as mines, smelters, power plants, and municipal waste treatment

new ──────────────▶ familiar

facilities release metals in to the environment. These heavy metals, especially cadmium,

new ────▶ familiar

then find their way in to the soil. The soil does not have an infinite capacity to absorb these

new ──────────▶ familiar

metals. Instead, unabsorbed metals move through the soil in to the ground water or are

new

extracted by crops that take the contamination into the food chain.

In measuring the length of your paragraphs, calculate size in final form, not in draft. Paragraphs that look like a good length on your screen may be too long when they appear on a web page or are set in narrow double columns of type (like those used in a newsletter or technical journal). They may be too short on a printed page of single-column type. Visualize what the paragraph will look like to the reader in its final medium.

Leading the Reader from Familiar to New Information

Start a paragraph with material that is familiar to the reader and move to the new. Note that approach in the two paragraphs shown in Figure 10.1, an original and a revised version of the opening of a report on methods for immobilizing cadmium in soil. The information in each is the same, but they differ in plan. The original immerses the reader immediately in new information—cadmium contamination—rather than beginning with the familiar and concrete—the sources of such contamination. The revision begins with the familiar.

Individual sentences in the revised paragraph in Figure 10.1 are also connected through a familiar-to-new structure. A subject familiar to the reader appears at the beginning, and the end tells the reader something new. This structure keeps readers from falling through the cracks of an argument and provides writers with an easy opening to a next sentence. You just rephrase the end of the one before, as in the following paragraphs (italics added):

In high-wage countries like the United States, industries stay competitive in a global marketplace through *innovation*. *Innovation* can lead to better production processes and better performing products.

Solar systems divide themselves into two categories: *active* and *passive*. An *active* solar system uses a mechanical pump or fan to transport heat from the collector to a *storage tank* for later use. While such *equipment* is more elaborate and costly than that for a passive system, it offers at least two distinct *advantages*. One *advantage* is its ability to *achieve hotter, more useful temperatures*. This *increased efficiency* leads to another advantage: the need for *less collector area*. The *collectors* fit more easily on to existing buildings than the bulkier passive equipment.

Showing Hierarchy or Parallelism

To help readers understand and remember, show the hierarchy from main points to subordinate, supporting points, and show the parallelism of equal items.

Hierarchy

The most-to-least-important pattern in the following paragraph provides supporting details for the generalization in the first sentence. The sentences are also connected in a pattern from familiar to new information (italics added).

> Wetland soils ("hydric soils") are usually deficient in oxygen and either permanently or periodically *saturated*. Such *saturation* can result from shallow ground water or surface water inundation. The degree of *saturation*—whether a soil is periodically or constantly saturated or is well drained—may be indicated by the soil's *color*. The *color* of a soil sample can be measured against the Munsell Soil Color Chart, which identifies color in a range based on *hue*, *value*, and *chroma*. *Hue*, which refers to the dominant spectral colors, is designated by a number from 0 to 10 and a letter indicating one or two adjacent spectral colors. *Value* indicates a range of shades from 0 (black) to 10 (white). *Chroma* indicates the relative purity of a color deviating from neutral, from 0 (neutral grey) to 20 (increased brightness). Thus a hydric soil color notation might be 10 YR 6/1: a hue of 10 yellow-red, a value of 6, and a chroma of 1.

The paragraph begins with a familiar term (wetland soils) and ends with the technical term (hydric), now known information.

Parallelism

The previous paragraph moves from a general statement about wetland soils to a specific measure of saturation, the soil's color. That mention of color leads to the description of a color chart, which identifies three qualities of color. Each quality is then defined as the subject of the three following sentences, in the order of their introduction. That expression of like ideas in like grammatical form is parallelism. Parallel construction has the added advantage of presenting new information in a familiar form.

In the following paragraph, the founder of the Center for Rural Design at the University of Minnesota supports his argument about the need to include the perspectives of both rural and urban residents in shaping future land uses. Both the first and the second sentences include parallel lists of items.

> Through design thinking and regional cooperation, communities found ways to produce goods and services that contribute greatly to economic development, environmental improvement, and quality of life. However, the design process must recognize the unique characteristics of each place, including its seasonal changes; its cultural, environmental, and social histories; and its practices for growing food, creating energy, and raising animals. The process must also identify opportunities for disadvantaged people, including Indigenous communities.
>
> (Thorbeck, 2020)

The first sentence lists three contributions, all logically equivalent as good effects and all nouns (development, improvement, and quality). The second sentence turns to a second set of conditions, called "unique characteristics," also expressed as nouns (with modifiers) introduced with the pronoun "its": seasonal changes, histories, and practices. The last sentence summarizes by repeating the term "process," not expressed as a term in the first sentence but included in the second.

Using Topic Sentences and Transitional Devices

You may preview in one paragraph the core idea of the next. The last sentence in the paragraph about regional cooperation previews the next topic to be considered: identifying the needs of a different community, signaled by the term "also." The sentence could instead be the topic sentence for the next paragraph if that need is to be further described.

You create coherence in a presentation by knitting paragraphs together through previews and reviews. You can keep a topic in mind without directly expressing it. But most of the time, you'll aid your reader by expressing the idea explicitly, up front, as the first sentence in a paragraph, which controls the evidence that follows. Note that process in this sequence of paragraphs about makerspaces (numbers are citations to sources). The sequence moves in a hierarchy from the most general definition (a physical location with fabrication and support) to particulars about these two offerings. Note, too, the familiar-to-new connections from sentence to sentence.

A makerspace offers a physical location with fabrication resources and support for students to learn and work in a hands-on environment [2]. More than just a fabrication facility, however, a makerspace aims to promote a sense of community. People matter just as much as (or more so than) machines. A participatory culture that encourages informal interactions among the communities served by the makerspace is what distinguishes it from a facility used only for fabrication.

The basic setting is an easily reconfigurable, relatively open, multi-use space. That means that much of the furniture and technology is on wheels or otherwise movable. While some space may be available for quiet, private work, most work is performed in the open. Team members can move chairs or stools around a table so that they can see what others are doing and learn from them. The space aims to foster a "shared environment for mistakes, inventions, and questions" [4, p. 1927]. Underlying the design of such spaces is a belief that design is "an inherently social process" [5].

The fabrication equipment and furnishing in the makerspace help "make thinking tangible," an important practice in interdisciplinary collaboration [3, p. 157]. The setting welcomes the mess that often accompanies innovation. A range of cheap materials, sometimes called "cruft," and 3-d printers allow for easy prototyping and iterative testing of design solutions. The making process is immersive and fluid. Devices for vertical and open display of writing and thinking abound. These include blackboards, whiteboards, folding screens or other tacking surfaces on wheels. There's an easy play between thinking, building a prototype, rethinking, erasing, reconfiguring. What's written on the boards establishes the team's common ground; the team sees each other's work as they fulfill mutual responsibilities to the project, building trust and gathering momentum and motivation.

Beginning every paragraph with a topic sentence may risk boring the reader. In most cases, take the risk. A strong topic sentence gives you a track to run on and your reader a way to skim and remember the document. Use the sentence to connect paragraphs, too, as in the previous paragraphs. Here are two other examples of topic sentences that mark a transition to a new topic:

- Having considered the disadvantages of Microsoft Meetings for workplace collaboration, we turn to its advantages.
- These statistics must be explained.

In addition, insert directional signals as appropriate to show turns in the discussion: "on the contrary," "despite these statistics," "considering this fact." Devices such as these, if used carefully rather than automatically, guide the reader within a paragraph and from paragraph to paragraph. Repeating key terms may also help: "This circumstance," "These devices," "Such ideas," "But these conditions." Such phrases or summaries further reinforce the pattern of familiar to new.

In a long discussion, you may devote an entire paragraph to reminding the reader of what has gone before and previewing what will come next. Such transitional paragraphs add no new information but mark the writer's kindness to readers:

This review of our company's environmental performance last year leads us to consider the changes in our own policies and the regulatory environment that will affect our performance in the coming year.

SENTENCES

A paragraph may consist of only one sentence, but usually a paragraph connects two or more sentences. Adjust sentence length to accommodate the reader, to segment information, and to show your personal taste. It's generally easier to go wrong in long sentences than short ones, so save long sentences for occasions when you know your reader is skillful in reading, and you feel in control of the writing. Don't fear short sentences. The period is one of the most underused—and desirable—marks of punctuation in scientific reports and articles. In general, avoid a succession of either long or short sentences. When you've written three long sentences, make the next one short.

As with paragraphs, you can take advantage of hierarchy and parallelism in the design of sentences, too.

Hierarchy
Design your sentence to emphasize the hierarchy of information when one idea is more important than another.

Flat: The condensed water then runs down the cover to a collection device, and the concentrated brines are disposed of.

Hierarchical: The condensed water then runs down the cover to a collection device that stores the concentrated brines for later disposal.

Much technical and scientific information reflects a hierarchy of cause and effect:

Because soils and rainfall vary, the soil moisture in individual fields must be monitored.

You may need to look beneath a string of modifiers to find the right connections or causal links:

Unclear connections: In Japan, some rice farmers use irrigation water contaminated with cadmium from mining operations, and the farmers show severe symptoms of cadmium poisoning.

Causal links: In Japan, rice farmers who use irrigation water contaminated with cadmium from mining operations show severe symptoms of cadmium poisoning.

A modifying clause at the beginning of a sentence produces anticipation and prepares the reader for something significant:

If we could once isolate the cause of the disease, then we could end the epidemic.

Dangling Modifiers

When you design information in a hierarchy, make sure you connect the modifying element clearly to the word it modifies. If not, you have created a dangling modifier, a phrase or clause that floats in a sentence. It either modifies nothing or attaches itself, wrongly, to the nearest available noun. When you want to be clear, attach the modifier. The introductory phrase in the following sentence dangles:

Dangling modifier: After reaching northern Alaska or the Arctic islands, breeding occurs in the low lands.

To correct, supply the proper subject:

Corrected: After reaching northern Alaska or the Arctic islands, the whistling swans breed in the low lands.

Or expand the participial phrase to a clause:

Corrected: After the whistling swans reach northern Alaska or the Arctic islands, they breed in the low lands.

Or rephrase the participle as a noun in a prepositional phrase:

Corrected: On their arrival in northern Alaska or the Arctic islands, the whistling swans breed in the low lands.

Ambiguous Modifiers

Avoid modifiers that look in two directions and thus cause ambiguity:

Ambiguous: I've been trying to place him under contract here for 3 years.

Does that mean you have been trying for 3 years, or his contract runs for 3 years? To indicate a 3-year contract:

Clear: I've been trying to place him under a 3-year contract.

Is the warning in the following sentence for car drivers about the trains or for the drivers of the trains about cars?

Ambiguous: The crossing signals do not provide a warning to drivers of approaching trains.
Clear: The crossing signals do not warn drivers that a train is approaching.

Inverted Subordination

Make sure the final structure of the sentence matches your intention. Problems occur when the subordination is inverted—that is, when main elements are expressed as modifiers:

Inverted subordination: Damage of surfaces by fretting is initiated by mechanical wear produced by the vibrating of the one surface on the other.

It's hard to know what's doing what in the sentence. The sequence of cause and effect becomes clearer when participial phrases are sorted into clauses with strong verbs:

Proper subordination: When one surface vibrates on another, this mechanical wear causes fretting, which damages the surfaces.

The following sentences also show how proper subordination helps clarify meaning:

Inverted subordination: The cut throat trout is a spring spawner over gravel pits in riffles of streams producing 3,000 to 6,000 eggs according to weight.
Proper subordination: The cut throat trout, which spawns in spring over gravel pits in riffles of streams, produces 3,000 to 6,000 eggs according to the trout's weight.
Inverted subordination: Cracking is a serious problem causing reduction of strength.
Proper subordination: A serious problem, cracking, reduces a material's strength.

Positive Phrasing

To avoid a buildup of modifiers, use positive phrasing. This advice, of course, applies to the grammar of your sentence, not its content. Readers can process positive statements more easily than negative ones.

Unclear negative: If the cooling system does not have sufficient coolant it will not be able to keep the engine running at a reasonable temperature.

Clearer positive: A proper amount of coolant will help keep the engine running at a reasonable temperature.

Unclear negative: When openness does not exist in an organization, many people are not kept informed of potential problems.

Clearer positive: Openness in an organization helps ensure that people are kept informed of potential problems.

Unclear negative: Do not send messages that cannot be made public.

Clearer positive: Send only messages that can be public.

Unclear negative: He did not neglect international issues.

Clearer positive: He accounts for international issues.

Parallelism

As in a paragraph, parallelism in a sentence or sentences reinforces the logic of a series and builds rhythmic expression. An *enumerator* term sets up the logic. The grammatical form of the first item governs the form of all the others (italics added to show parallelism in expression):

In the nineteenth century, proponents of domestic science *encouraged* proper training and education for its practitioners, *codified* standards of good practice, and *sought* status for professional housewives equal to that of other professionals, mostly men, in law, medicine, education, and engineering.

The next example lists roles (the enumerator term, not expressed) that the subject (he) performed, each expressed as a noun:

It was not as a surgeon, as the maker of his great museum, or even as a discoverer in science that he revealed his greatness.

Look for ways to straighten the parallelism of a rambling series:

Un-parallel: Overgrazing results in damage to the range, lower-quality livestock, and alters the numbers and distributions of other organisms, including small mammals.

Parallel: Overgrazing damages the range, lowers the quality of the livestock, and alters the numbers and distributions of other organisms, including small mammals.

In the next pair of sentences, you see how parallel construction hits a logical snag if the enumerator term doesn't work:

Un-parallel: The valving improvements we seek will increase reliability, accessibility, and maintenance and allow application to all sizes of valves.

Parallel: The valving improvements we seek will increase reliability and accessibility, decrease maintenance, and allow application to all sizes of valves.

The term *increase* is set up to control each of the following items. But it doesn't work for maintenance—presumably, the improvements should decrease maintenance. The second sentence corrects that problem in logic and brings the last item, "allow," more directly into the series. When you compare two items, tell the reader both halves.

Incomplete: The rotation span of an unmanaged forest is much longer than a managed one.

Complete: An unmanaged forest has a much longer rotation span than does a managed one.

Incomplete: Often farmers are able to plant their crops earlier in the spring because the field is dry.

Complete: If they adopt this tillage system rather than traditional techniques, farmers are often able to plant crops earlier in the spring because the field is dry.

Active and Passive Voice

Pay attention to verbs. Be particularly aware of whether you are using the active or passive voice. That's important, because the verb usually carries the sentence's main news.

Active: The team completed its prototype model.

Passive: The prototype model was completed by the team.

The active shows the subject of the sentence doing something. The passive shows the subject being acted upon. Although the active is often preferable, the passive may be necessary and effective in technical documents where there is no actor or agent (as in an automated process) or when the actor is unimportant:

- The sludge containing the calcium sulfate and calcium sulfite is pumped from the settling tanks to small ponds, where it is stabilized by the addition of fly ash and lime and then partially dried.
- The cancers are divided into three broad groups.

The passive also helps when you want to avoid accusing the person who is really at fault. A complaint manager might say, "The mower has not been lubricated according to warranty instructions," when someone complains about defects. It would be less tactful (if more concrete) to say, "You did not oil the machine when you should have." For years, using "I" or even "we" in technical and scientific documents seemed to deny the objectivity of research in the field, so editors and others discouraged it. Today, that prohibition has largely disappeared. For more about personal pronouns in technical articles, see Chapter 7.

LISTS

Within a sentence or paragraph, or as a unit in itself, use parallelism to create an enumerated list. Lists (like the one that follows):

- Segment information into small, easily digested elements.
- Sort and rank details better than running text.

- Increase the white space on a page or screen.
- Vary the look of a page or screen, especially in text that is otherwise dense.
- Reinforce the parallelism of a series or sequence.

Your thoughts may naturally come to you as elements of a list—for example, steps in a process, items in an inventory, lines in a budget, criteria for a system, advantages or disadvantages of a proposed activity. Or, in revising a draft, you may discover that running text in a paragraph contains a series of parallel elements that would become more accessible to the reader in list form.

Keep each element short, preferably one line. Order a sequence logically: by time, by priority, alphabetically by first letter, and the like. Introduce the element with a word (first, second) or number (1., 2.) that indicates ranking. Use an icon or bullet (•) for an unranked list. In general, omit punctuation at the end of each element unless it is a complete sentence in itself or completes a sentence introduced before the list.

Sometimes, short paragraphs are used as list elements, especially in the conclusion or recommendation section of a report. If you use this structure, however, make sure a visual cue or highlighting (for example, boldface on a key term that starts each paragraph) allows the reader to scan the page or screen to see the paragraphs as items in a list.

HEADINGS

In technical and professional documents, another unit of text, a heading, serves a critical role in showing the reader in advance the structure of the discussion. Headings identify the content, level of subordination, and relationship of document sections. The key words or phrases displayed as headings

- Ease skimming and foster remembering.
- Add an attractive design feature.
- Provide directional signals that help readers follow the turns of your presentation.
- Set up your document to be read by search engines or an AI system.

The box, "Key Words and Terms," provides more advice on this topic.

KEY WORDS AND TERMS

Key words are an essential point of connection between authors of journal articles and audiences interested in a particular topic who use search engines and AI apps to find relevant information. Compiling a list of key words is a good exercise, even if your ultimate goal isn't a published product.

Key words may derive from the technical terms that reflect how you and others in a discipline talk about your common knowledge. The terms provide a kind of shortcut for designating complex phenomena or concepts. Some researchers say that a discipline is essentially defined by its terms, which embed its view of reality. The terms also change over time in response to new developments in the field. So, especially in scientific and technical documents, avoid using synonyms. Make sure you refer to the same idea, event, mechanism, or organism with the appropriate technical terms and in the same way throughout a communication product.

More broadly, use terms that are culturally appropriate, particularly in reference to political entities and to human beings, tribes, ethnic groups, and the like. Terms carry emotional baggage. As an empathetic writer, use terms that your audience finds comfortable and understandable and that are ethically responsible.

For more about inclusive language, see Chapter 3

Title

The main heading is the document's title. Your reader notices it first, although you may consider it almost as an afterthought. Try thinking instead about composing a title early on. You then unpack and adjust the title as you develop the prototype draft. Make the title as simple and easy to remember as possible. Incorporate terms that would show up prominently in the search engines used by your target audience. Place those terms up front. Keep such terms as "final report" in the background.

Pay attention to the subject line of an email or memo. It's the title of the message. It alerts readers to the information that follows and to your purpose in sending, and theirs in reading, that message. Keep subject lines the same in a thread of messages. But change the subject line when you've moved onto another topic. It's easy not to. But down the road, you may have trouble retrieving an important message because it's hidden under the wrong subject line.

Directional Signals

Headings keep readers on track. Creating them also helps you impose an order on your discussion. The best headings occupy no more than one line, usually just a few words.

In a document-driven approach to structuring units of discussion, the headings derive from your outline, from your audience's explicit specifications, or from the conventions of the genre you are creating (or from all three). In a more design-oriented approach, you'll have more leeway with headings.

For more about collaborative writing, see Chapter 4

Levels

Headings denote the levels of discussion. In general, as you structure your discussion, avoid going beyond three levels of subordination. Group similar material under headings of similar rank. Rewrite to remove a single subheading (when you divide something, you come up with at least two parts).

Descriptive and Informative Types

Descriptive headings are pegged to the sections readers expect in any document. But within these general sections, or when your document is not constrained by set headings, choose informative headings. They help readers skim and rapidly retrieve the information they need. Note that all the headings in this chapter take an informative approach. Like many journal articles and technical reports, this chapter (and others in the book) avoids the heading "Introduction," on the assumption that the first unit of text is, of course, introductory.

Style

Keep headings parallel in logic and expression to reinforce their role as directional signals. Follow any specifications you're given by the person or organization you're addressing about heading style, including numbering, typography, and placement of headings on the page or screen. Whatever the form, be consistent. Take advantage, too, of the aesthetic possibilities in your headings, including graphic design, typography, and color.

PARAGRAPH TO PROTOTYPE

So far in this chapter, you have looked inside the unit of a paragraph to its components, including sentences and lists. With the section on headings, the discussion now turns in the other direction, moving outside the paragraph to picture the larger structure of a communication prototype composed of several or many paragraphs.

In the European tradition, a document meant to be read from start to finish often has three parts: introduction, middle, and ending or conclusion (see box on "Document Segments Aligned with Audience Expectations").

DOCUMENT SEGMENTS ALIGNED WITH AUDIENCE EXPECTATIONS

Introduction
> Engage the audience's attention.
> Set the scene: review the context for the communication.
> Create urgency: the purpose, the problem or request, why you're communicating *now*.
> State any limits in the scope or coverage, if needed.
> Define key terms, if needed.
> Establish the main point.
> Forecast the framework of the communication.

Middle
> Deliver the information you promised in modules that fit the framework you forecast.
> Periodically summarize what's been said and forecast what's to come.

Ending
> Provide closure.
> Remind the reader of the main point.
> Recommend future action, if appropriate.
> Tie up any loose ends while avoiding any new topics.
> Reinforce the audience's good impression of you.

Thinking in these broad terms can help you assemble your content for audiences expecting such a structure. In addition, apply some of the following guidelines as you outline and compose:

- Follow any stated request from the audience. As you've learned, much technical writing conforms to specifications—you write "to spec." This includes, for example, creating or updating statements required by regulatory agencies that document a company's compliance with policies and regulations. Because such updates are highly complex and legally scrutinized, structured authoring programs are often used to complete the task within highly determined constraints on text and visuals.

- Observe the structure associated with the genre of document you're creating—for example, a proposal, report, poster, email, or résumé. Genres change over time and are not internationally dictated, so adjust as needed to different forms in different cultures and communities. Apps, templates, and models are readily available online or in academic or corporate archives for the most common genres. Start, perhaps, with one of those applicable to your situation, but be ready to tailor it as needed.

  For more about these genres, see Part 4

- Similarly, take advantage of such time-tested general patterns for organizing content as hierarchy and parallelism. The patterns work in sentences, in paragraphs, and in whole documents. The traditional form of an outline reflects a pattern of hierarchy and parallelism, with indention helping you visualize the structure (Figure 10.2)

Other well-recognized patterns include comparison and contrast, elimination of alternatives, cause and effect, most to least important (and the reverse), or chronology. These patterns can be stretched to accommodate a variety of communication situations. You saw in Chapter 8, for example, how narrative, a form of chronological presentation, is becoming more significant in communicating technical and scientific information to the public. You'll need to be alert to new ways to structure text in new situations.

10.2
Hierarchy and Parallelism in a Traditional Outline According to the APA Style

Source: American Psychological Association, APA headings, Retrieved from https://bit.ly/3nnbMVG

First-Level Heading

First-level headings are centered at the top of new pages, in 14-point bold, title-cased (initial capital) letters, 1 inch from the top of the page, followed by a full line break (two returns).

Second-Level Headings
Flush left, 14-point bold, title case, preceded by a full line break, followed by a return.

Third-Level Headings
Flush left, 12-point bold italic, title case, preceded by a full line break, followed by a return.

   Fourth-Level Headings. Indented, 12-point bold, title case, ending with a period. Text begins on the same line and continue as a regular paragraph.

   Fifth-Level Headings. Indented, 12-point bold italic, title case, ending with a period. Text begins on the same line and continue as a regular paragraph.

Within the larger structure, you assemble sections, all supporting a main point. Adjust the size of the sections to the audience's motivation for reading, their capacity for paying attention, and the medium and mode of delivery. Determine what content is best expressed in text and what is best conveyed visually. Scientific researchers and medical professionals, for example, will slog through long passages of text, in a small font, online or in print. They'll also probably expect the dense, highly technical images you may have developed in your research.

Less motivated audiences need more breaks in the discussion and more devices aimed at guiding them through your argument, including shorter paragraphs, more transitional paragraphs, topic sentences for every paragraph, more headings, and attention-getting pictures. Keep in mind the principle that what the reader most needs must be most accessible to the reader. Simplify the structure where you can to save the reader's energy for understanding complicated material.

For more about simplifying the structure, see Chapter 11

In Part 4, you'll find detailed guidelines for structuring text in the many communication situations you'll encounter as a technical and professional communicator. Some of these situations are relatively routine, and some will require more customized approaches. Your skills as a communicator depend in part on knowing the difference, even when something seemingly routine calls for a second look. Design thinking will give you an edge in making the right choices as you write and structure content to meet your reader's purpose as well as your own purpose in communicating.

This chapter has reviewed the basic units of text—paragraphs, sentences, lists, and headings—you will use in composing a prototype draft. With paragraphs, you adjust your content in modules that match the capacity of the reader to understand and pay attention. Paragraphs both stand still in being devoted to one topic or subtopic and move, in bringing the reader from familiar information to new information. The other basic units perform their roles within paragraphs to make information readable. You assemble paragraphs into segments that work together within a document whose structure is designed to meet your reader's expectations. That framework may be one common in many documents, a pattern of introduction, middle, and conclusion. It may be dictated by a reader's explicit request or the segments expected in a particular genre. Or, enhanced by design thinking, you may develop new frameworks or formats to meet new situations for communicating.

CHECKLIST: COMPOSING AND STRUCTURING TEXT

1. In paragraphs and sentences
   Lead the reader from familiar to new information
   Show hierarchy or parallelism explicitly
   Adjust the length of paragraphs and sentences to the audience's capacity
       for attention

2. Use enumerated lists to
>   Reinforce the parallelism in a series
>   Increase white space on a page or screen
>   Ease the reader's task in understanding and remembering discrete items
>   Break up an overly text-based communication

3. Use headings to
>   Keep you on track in writing and help your reader skim to find points of interest
>   Provide key words for automated information retrieval
>   In a subject line, alert readers to the topic of an email or memo

4. Structure the text as a whole
>   Assemble units to meet a reader's direct request
>   Align units in the common pattern of introduction, middle, and ending
>   Conform with the structure of the appropriate genre for the situation
>   Through design thinking, create custom formats and structures for new situations

---

EXERCISES

1. The following paragraph provides confusing signals to readers about what they are expected to know and what material is expected to be new to them. It also announces topics it doesn't really treat in the way announced. Revise the paragraph to straighten its approach to the reader.

   - *Foam applications* is the phrase used to describe the process which applies additives to a moving paper web as foam. This process has many advantages over conventional liquid applications. Two problems have hindered the progress of this new technology. The process must be established in the paper industry. The speed limits encountered in the current application technology must be increased in order to expand the number of paper machines which can potentially use foam. The development work run at the Placerville plant is attacking both of these problem areas and this report summarizes the achievements which have been accomplished there during the year.

2. Here are some redundant expressions common in technical prose. Discuss why the phrases are redundant and add any similar phrases you find in your reading. Keep this list handy as you write so you avoid such expressions:

   - completely eliminate
   - collaborate together
   - few in number

- estimated at about
- component part
- background experience
- past history
- oval in shape
- hot in temperature
- red in color
- they are both alike
- adequate enough
- a funding level of $100,000 in magnitude
- final end
- cheaper in cost
- true facts
- close proximity.

3. Technical and professional writers often smother their verbs by transforming action into a noun or by using unnecessarily long verbal phrases. Such usage diminishes accuracy and bores readers. The following pairs show how you can clarify verbs:

- *Smothered*:

  o make contact with
  o make a purchase, give approval to, have a deleterious effect on
  o have a tendency to
  o have an influencing effect on

- *Clear*:

  o call, write, visit
  o buy, purchase, approve, harm
  o tend to
  o influence, affect

Good verbs give energy to sentences. Note how transforming a noun back to a verb energizes the corrected sentence in each of these pairs:

- *Transformed verb*: Explanation of the variables is included in the report.
- *Strong verb*: The report explains the variables.
- *Transformed verb*: Camouflaging the vehicle is the next step you should perform.
- *Strong verb*: Next, camouflage the vehicle.

Unsmother the verbs in the following sentences:

- Mating of grizzlies takes place in the month of June.
- There was a variation in weight.

- A review of our customers' needs for such compounds was undertaken.

4. Convert the following passive sentences into the active voice.

   - Control of distortion can be accomplished with the following machines.
   - The solution of this problem has been obtained by adding sufficient deoxidant.
   - As soon as the explorers reached the ice, camp was set up.
   - An improvement in quality has been made.
   - The values are found to be in agreement.

5. Correct errors in parallelism and incomplete comparisons in the following sentences:

   - Seawater may consist of whatever type of water the ship is operating on—fresh, brackish, or seawater.
   - Factors contributing to increased pollution included population increase, industrial expansion and their waste, pesticides poisoning wildlife, and people who used their trucks and automobiles more.
   - High nutrient levels in our waterways result from deforestation, mining, farming, and from the manufacturing segment of our economy.

6. Straighten the hierarchy in the following sentences so that main ideas stand out from their supporting evidence:

   - It is important to mention, when talking about losing weight, that changing a diet should not be the only way of losing weight, changing a diet with an appropriate daily exercise regimen.
   - There are currently two bills being considered by Congress that intend to enhance the vaccination program for infants in rural areas and inner cities.

7. Make a list of any errors in paragraph and sentence structure as well as in conventions of grammar cited by your instructor in your assignments. Review the list with each assignment. Try to make sure that no item appears more than once—and that no additions are necessary by the end of the term. Enter search terms such as *grammar* and *style* into a search engine to retrieve extensive advice about usage.

FOR COLLABORATION

Assume your team of two or three has come up with the following items of information after a brainstorming session about the Delaware Estuary. You want to include these in an introductory section to your report on a proposed regional program that will institutionalize wise management and conservation of the estuary. Develop a central point for your description and a plan for structuring the information. At your instructor's request, write the introductory paragraph or concluding paragraph—or entire description.

- Over 90 percent of the area meets the "swimmable and fishable" criteria of the Clean Water Act
- Definition of an estuary (transitional zone where salt water from sea mixes with fresh water from the land)
- Much cleaner today than ever
- Major problem: the reach that flows between Philadelphia and Camden has lots of bacteria and a higher than acceptable level of dissolved oxygen
- Productive habitat
- Breeding, spawning, and feeding grounds for fish
- Migratory waterfowl, reptiles, mammals live there
- Fishery is down since 1950s
- Buffer upland areas from flooding
- Overfishing has reduced levels of estuarine-dependent species
- Greenway trails being developed to connect historic sites and recreational facilities
- Habitat destruction and lowered water quality led to decline in fishery, so shad, Atlantic sturgeon, and striped bass are below historic levels, although some improvement
- Estuary naturally protects drinking water—filters pollutants and sediment
- Heavy use of water for industry and homes threatens long-term supply
- Some fish die when they get caught in power plant water intakes
- Show all the states included in the estuary program area (map?)
- Toxic substances are at elevated levels in sediment and in the organisms that live in the estuary, so there have been fish consumption advisories
- New public parks in the area
- Show the three zones (upper, transition, lower)
- The fragmentation of the habitat, especially loss of freshwater wetlands, means that organisms which need specific types of habitats aren't thriving
- New forms of shipping have led to new commercial uses and reinvigorated urban waterfronts
- Deterioration of the estuary's ability to buffer the impact of pollutants
- Poor land management means that a lot of agricultural land and natural habitat is eaten up with houses.

REFERENCE

Thorbeck, D. (2020). Shaping futures through land use and design. *Architect*. October 28. Retrieved from https://bit.ly/2QocupC

11 Revising the Design

A popular brand of beer in the US used this tag line in its ads: "When it's right, you'll know it." The motto works for communication products as well. When your message is right, your *audience* will know it. How do you know when your prototype draft has become the right product? In this chapter, you will learn strategies for testing, evaluating, and giving final polish to a draft. The process is called *revising*, *reviewing*, or *editing*. These terms apply to subtly different aspects of the process. But all refer to ways of looking back at what you have created to make sure it works.

A relatively linear view of the writing process sees revising as a final step. The design-thinking approach you're learning in this text is different. Revising is integral to the cycle of creating a draft, testing, evaluating, and redesigning. You revise in iterations as you enhance your understanding of your project, of your audience, and of strategies needed to communicate the project's results to your audience. Because much composing in projects is social writing—that is, done in the open with others, this chapter also provides guidelines for asking others to look at your work as you, too, give effective advice.

For more about collaboration, see Chapter 4

TAKING A SNAPSHOT

For more about headings, see Chapter 10

Start each phase of the revising process with a quick snapshot of the draft. Focus just on the headings. That means, of course, that there are headings, either derived from an outline you created before writing or inserted while you wrote a draft. Test a draft by listing headings separately from the text. Write the list on a whiteboard or other device where you can share the list with your team.

The list of headings shows at a glance how the units of your discussion support a main idea. It can reveal, for example, problems in emphasis. An important idea, covered in the text but not noted in a heading, may miss its mark. The list may also reveal digressions from the main idea, from the specifications given to you by the audience, or from the conventions of the genre. It may reveal omissions in information. You make mid-course corrections to get back on track. At each phase in testing your prototype draft, review the headings to make sure they designate the right content and signal the right level of subordination. Let the list of headings spark a discussion about redesigning the structure to make the information most needed by the audience most available to them.

Answer these questions about your headings:

- Does all information under one heading really belong there?
    - o Is the relative emphasis indicated by the headings appropriate?
    - o Are there enough headings to be meaningful?
    - o Does an excess of headings dissolve emphasis and continuity?
- If you have more than three levels of headings, how can you reorganize to reduce the levels?

If you have a single subhead, restructure the section.

TESTING AND EVALUATING INFORMATION

When you spot problems in your draft, sort through their possible causes. One cause may be that your information is not right. Your information should meet five standards. It should be comprehensible, credible, timely, adequate, and relevant. Here are some ways to test whether you are meeting those criteria.

Your information is *comprehensible* when the audience is capable of understanding it. Avoid unnecessary complexity. Use information that enlightens rather than obscures, and don't try to impress your audience with fancy statistics or bizarre facts.

Your information is *credible* when it derives from careful observation or accepted authorities (including yourself) and can be independently verified.

For more about credibility, see Chapters 6 and 7

Your information is *timely* when it reaches the audience when the audience needs it. You provide the latest evidence for decision-making, for example, when the audience is ready to make that decision.

Your information is *adequate* when it serves the purpose that you and your audience share in the communication product. Provide enough information to support your argument or analysis.

Your information is *relevant* when it relates directly to the matter at hand. You may be naturally inclined to include whatever information you spent time and energy developing. But relevance needs to be the test. Your information is the wrong information if it's not what the audience asked for or needs.

Explain any anomalies in your information—for example, unexpected patterns, data that don't appear to make sense, odd spikes. If your audience might question an issue, anticipate that question. Make notes as you carry out the review about gaps or questionable information in your draft to be addressed in the next draft.

TESTING AND EVALUATING CONTENT AND STRUCTURE

Another cause of problems in a draft, in addition to having the wrong information, may be that your content is not expressed appropriately, in text and visuals, for the audience. A third cause may be that the structure isn't right.

For more about composing visuals and text, see Chapters 9 and 10

The box on "Checklist for Testing and Evaluating a Communication Product" summarizes a three-level strategy for evaluating content and structure. At the first (top) level, you measure the draft's *effectiveness*. Effectiveness is doing the right thing. At the second level, you measure the draft's *efficiency*. Efficiency means doing things right. When you assess the effectiveness of a draft communication product, you focus on the big picture. Does this email, report, proposal (or other product) accomplish what I want for its intended audience, with the right evidence? When you measure efficiency, you test whether you're using the best, most direct expression, structure, and design. At a final review, you also match your draft against the standards or conventions of language and usage expected in your culture, business, and particular company. A review of the conventions includes attention to the details of grammar, punctuation, and style that your audience assumes.

CHECKLIST FOR TESTING AND EVALUATING A COMMUNICATION PRODUCT

Effectiveness: Begin at the Top

- Start by answering the audience's question: So what?
- What is the main point?
- Is that the *right* point, now that you have had time to reconsider it?
- Is it obvious and clear?
- Are counterarguments or alternative approaches properly dealt with?
- Will the audience know what to do next—what you expect of them?
- Distinguish the new from the familiar, and start with the familiar.
- Build modules of support for the main point, asking and answering questions, making and defending claims, providing explanations and definitions.

Efficiency: Check Content and Structure

- When you're satisfied with the big picture, check your content and structure.
- If you anticipate resistance to your main point, show the audience explicitly how the subject affects them.
- Signal relationships in your content with forecast statements, headings, connecting words, repetition, layout, type, color.
- Make smart choices between visual and textual content.
- Place major statements prominently, in headings or topic sentences.
- Check that statements in the introduction and conclusion match.
- Provide explanatory background if your main point will cause difficulty.
- Identify places where you need to extend the discussion of main points with definitions and subpoints.
- Look for ways to shorten the text by deleting unnecessary details.
- Eliminate statements that raise unnecessary questions.
- Divide long paragraphs into smaller ones.

Conventionality: Meet Expected Conventions
- Adjust the level of language (word choice) to the audience.
- Conform to conventions and any applicable style guide.
- Focus on sentence construction.
- Review all verbs—make them work, keep them precise.
- Proofread for spelling, punctuation, mechanics.
- Proofread again.

REVIEWING COLLABORATIVELY

You've probably been testing and evaluating drafts of documents for years. You may have done so by yourself or with help from others. Collaborative projects depend on effective practices for commenting on each other's communications. This situation presents many opportunities to offend and to be offended. That's one reason why it's best to do so as part of an iterative process, especially if the prototype draft is visible to the group on a whiteboard or a screen. You learn from each other as you assess what works and what doesn't and make changes while the group talks. Everyone sees the changes as they are implemented. You work in common to achieve the best common goal.

As you move in toward a final prototype communication product, you may share and comment on documents through Google Docs or your workplace collaboration software. Each team member contributes comments, sometimes called "annotating," separately. Depending on the team's agreed operating style, you may all be able to make changes directly in an assembled text or text sections through a reviewing function. Or you may need to contribute to a separate thread of comments.

Confirm the Level of Review Sought

Writers asking for help should first clarify what kind of help they want. But if you're asked to simply "look over the draft," you might suggest possibilities in terms of specific questions. To avoid offending the writer, and wasting time crafting comments that aren't wanted, determine how invasive the writer wants you to be:

- Proofread only? Look for simple errors in punctuation and spelling and ignore the rest.
- Restructure? If the writer hasn't made a list of headings, make such a list and discuss whether that outline conforms with the writer's intentions.
- Offer suggestions for a particularly sticky section? Confine yourself to that section.

The three-level approach shown in the box "Checklist for Testing and Evaluating a Communication Product" provides a good way to frame questions about what a writer wants and how you should respond. You'll save on hurt feelings and wasted time by determining whether the writer really wants your serious feedback or is simply looking for a thumbs-up comment in a thread. A mere thumbs-up may have ethical considerations if you're ignoring a problem. But that's another matter. Your more extensive comments are not likely to have the desired effect if the writer ignores them.

If the writer is serious about learning what might need to be changed, then review the text along the lines suggested below.

Praise Good Work

Cite places where the writer handled material particularly well. Praise is a strong motivator. In addition, sometimes writers don't really know what they've done well or badly. Reinforcing their knowledge of what's good helps them not to change that and to imitate their good skills elsewhere.

Assess Your Standing as a Reviewer

For example, if you are the designated writer and editor on the team, your word goes. In assigning you that task, the team has also given you final responsibility for the document. If everyone on the team is a peer, then your word is equal in weight to the others. You're jointly responsible for reviewing the product at each level, for representing the purpose and perspectives of the client or sponsor, and for measuring others' writing against the team's goals. If you are the copyeditor, your task, the final one before you hit "send," is to measure the text against preset and primarily objective standards of format, mechanics, usage, and style.

Your role determines the weight of your comments and the amount of discretion the writer has in accepting them. Note that none of these roles is "English teacher." You're not grading the writer: you're helping to ensure the quality of the final product.

Take Your Role Seriously

Take other writers seriously. Be sensitive to their feelings. Such sensitivity is particularly important when you deal with people who might be embarrassed about their writing skills, who are writing in their second (or third or fourth) language, or who may have been brought up to avoid open criticism and confrontation.

Control the Impulse to Rewrite

Commenting and editing are repair work, not construction. If you rewrite, you risk misunderstanding the writer's point and changing the substance. You may also inadvertently introduce a new style or tone that clashes with the original. Modify the writer's work only to prevent real damage. You want to interfere as little as possible.

Be Specific

Start with the big picture and work toward a word-by-word analysis (unless you are a copyeditor whose major interest lies in the wordby-word analysis). Write a summary of your criticisms for a long report. Use consistent notations.

Don't Nitpick

The writer has only so much energy for revision. Don't waste that energy by focusing on a few minor problems in comma usage on page 1 when, for example, a lengthy discussion on page 12 should appear as a table rather than text. Just as hospital emergency rooms sequence patients for care based on the acuteness of their condition, practice a kind of triage on the writing you're reviewing. Distinguish between preferences and requirements. You may think "utilize" sounds better than "use," but generally the words are interchangeable. Avoid changing the writer's style merely because it isn't yours. Achieving a coherent voice in a document often necessitates the team's agreement on matters of vocabulary, including key words. Rely for reference on your organization's style guide and the conventions of your discipline. You may need to explicitly discuss terms that are appropriate and ethical in the setting for your communication.

Especially on an international team, creating an appropriate voice may be difficult. An American, for example, may see as praise the characterization of her prose as "hard hitting." Someone from a less confrontational culture might find that a negative. Vague or indirect language may represent the writer's polite stance toward the reader, but it may also confuse the reader.

REVIEWING THE FINAL DESIGN

In the final review, step back to see the communication product as a whole. Consider the pages and screens you have produced as visuals themselves that in turn embed textual and visual segments. In Chapter 7, you read guidelines for investigating external sources both to glean information and to use them as models for your own writing. Look at these sources as well for their *design*. What do their pages and screens *look like*, apart from what they say? Pay attention to the design of their print and digital surfaces. Collect models of good design to enhance your own strategies.

For more about digital surfaces, see Chapter 8

This section provides guidelines for a final review of the content, structure, and design of your communication product. In this review, aim to

- Establish an appropriate persona.
- Simplify.
- Clearly signal the relationships among segments.
- Pay attention to how the audience/user will experience your product.

Establishing an Appropriate Persona

Your persona reflects your relationship with and attitude toward your audience. It comes through in your style and design. You ensure that the content of your communication is as easy as possible to understand and act upon. At the same time, you establish your credibility as someone who can be trusted. It takes trial and error to find your voice, in words and images, which is just what design thinking enables.

11.1
Two Paragraphs
from a Scholarly
Publication

Source: Davis
(2020, p. 10)

> But looked at in its totality, like any organism, the city includes things that can be seen and things that cannot, things that are pleasant and fun and things that are not, the 'stage' as well as the 'back of the house'. And even if production takes place outside the boundary of the city, 10,000 miles away, this points up the difference between the functional boundary of the city and its political boundary. The city needs to be understood as a functionally-whole entity, not artificially severed at its political boundary. It is the functional completeness of the system that should determine its geographic definition and not the other way around--and this itself helps provide the argument for the inclusion of production in our full understanding of the city.
>
> The city is many things at once: a linked system of changing human communities; a physical artifact made up of other physical artifacts; an economic machine; a dense system of communication; a mechanism for cultural maintenance, innovation, and transmission; a framework for a bioecological system of insects, plants, birds, and animals; a place where people can live closely with each other and where children are socialized. All these things operate together and at different scales, ranging from the city as a whole in its interaction with other cities, to an electrician buying a coil of wire at the electrical supply house for the installation of a walk-in refrigerator in a restaurant kitchen.

One way to craft your own style is to imitate a professional style you like. Your collection of models of good document design will help. Figure 11.1 suggests some dimensions of a professional persona. It provides two paragraphs from a section headed "The nature of cities" in a book, *Working Cities*, whose author, a professor of architecture, argues that studies of cities have overlooked their role as centers of production, not just of "consumer culture." The paragraph before these two paragraphs ends: "The idea of the city as a 'Disney-fied' semblance of itself has been written about extensively."

In beginning "But," the first sentence signals a change in the direction of the argument from the notion of a city that has been "written about extensively," the phrase ending the previous paragraph. The new concept, to be developed in this scholarly book, is presented by analogy: the city as an "organism." A second analogy, to help the audience understand the author's concept, evokes a dramatic performance, the stage, and back-of-the-house, terms that connect to the Disney-fied mention above. The author, however, points to "things that cannot be seen" by visitors to an amusement park but can be seen in a city. In the next sentence, beginning "And," the author further supports his argument about "functional completeness," as opposed to more arbitrary political boundaries, that should define a city geographically. The three repetitions of the term "functional" emphasize this organic unity.

The second paragraph further details unity, because the city is "many things at once." The linked characteristics, parallel in expression and all in one sentence ("at once"), support this point. The list moves from the abstract, "bioecological system," to the specific, "insects." The second sentence begins with a summary of these characteristics ("All these things"), reinforces the notion of scales, and ends with a vivid, concrete picture of an electrician buying wire. The paragraphs are relatively short and tightly connected.

If you're used to reading textbooks aimed at instruction, or short bits of text on a screen, you might find that first sentence in the second paragraph off-putting. It could be recast:

The city is many things at once:

- a linked system of changing human communities
- a physical artifact made up of other physical artifacts
- an economic machine
- a dense system of communication
- a mechanism for cultural maintenance, innovation, and transmission
- a framework for a bioecological system of insects, plants, birds, and animals
- a place where people can live closely with each other and where children are socialized.

You'd expect the sections of a textbook following this sentence to treat each of these topics in order, with key terms as headings. That list could set up the structure of an entire document. It would help you learn and remember separate units. But it lacks the momentum, the at-once argument central to the author's concept of the city. The two approaches to the same content speak differently and represent a different relationship between author and audience.

Simplifying

To follow the guidelines in the box on "Checklist for Testing and Evaluating a Communication Product" and, especially, to make the information the audience most needs most available to the audience, simplify your communication product. Many designers note that a major role of design is to bring order out of chaos. The more complex the information, the simpler your design should be. Look for the underlying bones of an argument. Establish and repeat key elements rather than distracting the reader with unneeded variety.

For more about visuals and graphic design, see Chapter 9

The two paragraphs concerning the disposal of aerosol cans (see the box) show how an original can be revised with those guidelines in mind. In particular, the revised version brings out a main point buried in the original list. It establishes a hierarchy. Boldface on that point adds to the emphasis. The original list is jumbled. The revised list also provides steps in a process, in order, parallel in expression, each starting with an imperative verb.

ORIGINAL

In short, to properly recycle used aerosol cans as a scrap metal, you should:

- Designate a container for the collection of used aerosol cans at your business;
- Label the container "Used (Empty) Aerosol Cans Only";

- Remove the plastic spray knob from each can. Place the plastic spray knob in the trash;
- Use the designated recycling centers to recycle your used aerosol cans (yellow recycling containers);
- Never dispose of used aerosol cans in the trash. You could be illegally disposing of a hazardous waste which may result in enforcement actions and accompanying penalties;
- Take no more than 50 used aerosol cans at a time;
- In addition to empty aerosol cans, you can also recycle aluminum and steel cans with the DSWA; and
- Items <u>not</u> accepted through this metal recycling program include: foil or pie trays, paint cans, propane cylinders, siding, beach or lawn chairs.

REVISED

Never dispose of used aerosol cans in the trash. You could be illegally disposing of a hazardous waste, which may result in enforcement actions and accompanying penalties.

Steps to properly recycle used aerosol cans as scrap metal:

- Designate a container at your business for the collection of used aerosol cans
- Label the container "Used (Empty) Aerosol Cans Only"
- Remove the plastic spray knob from each can before throwing it into the container
- Bring your cans (no more than 50 at a time) to the closest recycling facility and place them in the designated yellow containers.

In addition to empty aerosol cans, the designated recycling center also accepts aluminum and steel cans for recycling. Items **not** accepted through this metal recycling program include foil or pie trays, paint cans, propane cylinders, siding, beach or lawn chairs.

Signaling Relationships among Segments

In Chapter 10, you learned strategies for composing and assembling units of text—sentences, paragraphs, lists, and headings—to match your content to your audience. In the final design review, ensure that you have also used graphic devices strategically to signal the relationship among the units. These devices include typography, color, and layout.

Changes in Type

For more about typography, see Chapter 9

You can signal a change in the level of content with a change in typeface, font, or size. A variation in type size within a document signals that the unit in that type is either more (if larger) or less (if smaller) important than the rest of the text. For example, cautions and warnings are often set in larger type, sometimes accompanied by a graphic symbol. Footnotes are often in smaller type. To describe all the potential side effects and cautions concerning use of prescription drugs and meet with regulatory agency guidelines for disclosure, companies often use a very small type size on folded inserts packaged with the drug. Some "fine print" in agreements for the use of software and contracts raises ethical issues when it deliberately downplays significant information.

For running text in whatever font you choose, always blend upper and lower case letters. Text in all capital letters draws attention, but readers quickly tire of focusing on only one word at a time. The practice also carries connotations of shouting. Use boldface, italics, or other highlighting to make some text, such as headings, stand out from supporting text. Make sure your usage is consistent. In general, keep text variations within one, or possibly two, font families.

To signal the relationship among headings:

- Use type sizes from 12 point to 36 point. The size shouldn't be smaller than the text it introduces.
- Develop a consistent system to indicate the heading level through typography, numbering (if expected by the audience), and placement on the page or screen. A reader should be able to quickly recognize main headings.
- Consider not capitalizing the initial letters of second and subsequent words in a heading for a more modern look and faster reading.
- Avoid punctuation (colon or period) at the end of a heading. The mark subtly stops the reader rather than inviting them into the text.

Color

Use color or shades of gray to show hierarchy or parallelism in segments of the text. A major statement highlighted in color stands out. Warnings can be signaled by red type or by a red background displayed consistently across several pages of a document. Color edge strips in a printed document can help readers locate content rapidly. For example, a report on a proposed interstate highway identified the alternatives by color: a yellow route, a red route, and a blue route. Sometimes, the labels stick for years. Interstate Highway 476 around Philadelphia is known to long-time locals and radio traffic patrols as the "Blue Route," which was its name in preliminary proposals.

Other Graphic Elements

Graphic symbols and signs can also be used to designate units and their relationship, as well as to attract audience attention. For example, use a *rule*, or line, to divide the page either vertically or horizontally. Rules can box either text or visuals. Page numbers can also become graphic features. Incorporate icons for further visual interest, as in Figure 11.2.

Layout

To achieve the right look of the whole document, place visuals, text, and other elements on a page or screen in a way that helps readers understand the relationships among the units. That placement is your *layout*. Begin by determining the area you have to deal with—that is, available size of a page or screen area. Will the reader see one page at a time or a two(or more)-page spread? How much of the screen will be occupied by fixed elements, such as tool bars? The standard page size in the US is 8½" × 11". In the rest of the world, most documents are printed on A4 paper (210 mm × 297 mm). You'll need to adjust text area accordingly.

Effective layout is based on two strategies: *alignment* and *grouping*. As an example, look at Figure 11.2, which reproduces a page from the *AIA Framework for Design Excellence*. This publication is also available online. Produced by the American Institute of Architects, the framework "represents the defining principles of good design in the 21st Century":

> Comprised of 10 principles and accompanied by searching questions, the Framework seeks to inform progress toward a zero-carbon, equitable, resilient, and healthy built environment. These are to be thoughtfully considered by designer and client at the initiation of every project and incorporated into the work as appropriate to the project scope.
>
> (AIA, 2020, p. 1)

The document "challenges architects with a vision of what the profession strives to achieve" and "provides practical resources to help all architects achieve the vision."

Note the horizontal rule at the top of the page and the icons next to each of the units. The units are not numbered—probably a deliberate choice to emphasize the equality in importance of each guideline. Numbers tend to indicate steps or an order of importance. The headings for all units are parallel (beginning with "Design"). They are *aligned* on the page along an imaginary vertical line, off center and equally distanced from the identifying icons, which are also drawn to the same scale and aligned in a column in the wide left margin. Each unit is a group consisting of similar elements: the icon, the heading, a brief statement beginning "Good design + verb" or "Design solutions + verb" to indicate what design *does*, not just *is*. What follows in each unit is an indented and bulleted list of specific questions that designers should answer as practical resources to "achieve the vision."

Significant white space surrounds the units to achieve a clean look and easy readability. Note that the text is *left justified*—that is, there's a ragged right margin. That ragged edge helps the reader's eye mark its place before returning to the even left margin. Published text (like this textbook) tends to be fully justified— that is, adjusted so that every line is the same length, aligned at both the left and right margins.

Paying Attention to User Experience (UX)

Whether you are creating a data chart, research proposal, recommendation report, or instructions for the myriad of situations you encounter in the workplace, you should apply design thinking in the invention *as well as* the revision process. Indeed, design-thinking practices can help you create user/audience-centered communication products. In the revising process of a product and its continuous

Design for integration

Good design elevates any project, no matter how small, with a thoughtful process that delivers both beauty and function in balance. It is the element that binds all the principles together with a big idea.

- What is the concept or purpose behind this project, and how will the priorities within the nine other principles inform the unique approach to this project?
- How will the project engage the senses and connect people to place?
- What makes the project one that people will fight to preserve?
- What design strategies can provide multiple benefits across the triple bottom line of social, economic, and environmental value?

Design for equitable communities

Design solutions affect more than the client and current occupants. Good design positively impacts future occupants and the larger community.

- What is the project's greater reach? How could this project contribute to creating a diverse, accessible, walkable, just, human-scaled community?
- Who might this project be forgetting? How can the design process and outcome remove barriers and promote inclusion and social equity, particularly with respect to vulnerable communities?
- What opportunities exist in this project to include, engage, and promote human connection?
- How can the design support health and resilience for the community during times of need or during emergencies?

Design for ecosystems

Good design mutually benefits human and nonhuman inhabitants.

- How can the design support the ecological health of its place over time?
- How can the design help users become more aware and connected with the project's place and regional ecosystem?
- How can the design build resilience while reducing maintenance?
- How is the project supporting regional habitat restoration?

AIA Framework for Design Excellence   2

11.2
A Page from the
*AIA Framework for
Design Excellence*

Source: Reproduced with permission of the American Institute of Architects, 1735 New York Avenue, NW, Washington, DC, 20006

iterations, you should evaluate how the communication product meets (or misses) the desired results by studying the actual users' experience through testing and user research.

Product testing can take many forms. You can learn about popular methods in Chapter 17, where you practice testing instructional materials and gathering user insights. Beyond the usability (efficacy and accuracy) of a product, user experience is concerned with the visceral/immediate and sustained reactions of a user to a product. Among the chief questions you may ask during the revision process that focus on UX are:

- What do users find most enjoyable about the communication product?
- What do users find most frustrating about the product?
- If they could change one thing about the product, what would it be and why?
- Which features of the product could the users live without?
- Which features could they *not* live without?
- Overall, how easy to use do the users find the product?

The goal of user-centered design in technical communication is to empathize with users through understanding their needs and desires. These things can be easily overlooked if you only focus on the functionality and forms of your product. Many technical professionals have advocated for greater care in communication design following emotional design principles (e.g., Norman, 2004; Walter, 2011) and going beyond usability and reliability.

When you write a quick comment or simple email to a friend, you can probably get away with drafting and sending it without much looking back at the text you've sent. When the situation entails more significant consequences for you and your audience, look back. Look back in a continuous cycle while you draft and revise. Compose text and visuals, review, identify problems, refine criteria, compose again. In this chapter, you've learned design-thinking-oriented strategies for revising. After taking a snapshot of where you are, you test and evaluate different dimensions of your draft. You test your information. You test the effectiveness, efficiency, and adherence to conventions of your product. You confirm that all the parts fit together in a pleasing and accessible final design. You see the final page or screen as a visual in itself. In this chapter, you've also learned strategies for talking about your prototype with others, including paying attention to how a user will experience your product.

CHECKLIST: REVISING THE DESIGN

1. Review and revise your prototype draft iteratively as you compose
   Take a snapshot of a draft by listing and reviewing headings
   Make sure you have included all essential and appropriate key words
   Make sure your information is comprehensible, credible, timely, adequate, and relevant

2. Test and evaluate your content and structure
   Test for effectiveness
      Answer the audience's key question: So what?
      Focus on a main point—the right one
      Make the point obvious and clear
      Ensure compliance with any request or genre specifications

    Test for efficiency
        Sort new from familiar content and move from familiar to new
        Distribute information strategically between text and visuals
        Help the reader find, understand, and remember your message
        Divide long paragraphs into shorter ones

    Meet conventions of grammar and style

3. Comment appropriately on others' writing
    Praise good work and be polite
    Determine the extent of commentary desired by the writer
    Assess your standing as a reviewer
    Control the impulse to rewrite

4. Complete a final design review
    Establish an appropriate persona
    Simplify
    Clearly signal the relationships among segments
    Investigate how the audience/user will experience your product

---

EXERCISES

1. Identify the main point in the following paragraph and reorganize the sentences to support it in a coherent unit. Hint: underline "Hauser Lake Dam" each time it appears and line up information around it.

   (1) Around 1900 an attempt was made to use steel as a major construction element in three large dams. (2) However, after the failure of the Hauser Lake Dam, the idea was generally abandoned. (3) Today steel dams are used only as temporary cofferdams needed for construction of permanent dams. (4) Basically a framework covered with riveted steel plate is used for the steel dam. (5) A masonry abutment is used to anchor the steel work into the reservoir. (6) A typical design is shown in Figure 19 [not reproduced here]. (7) The Hauser Lake Dam in Montana failed on April 14, 1908. (8) Steel dams decreased in popularity after this. (9) There were only three major steel dams made. (10) These were the aforementioned Hauser Lake Dam, the Red Ridge Dam in Michigan and the Ash Fork Dam in Arizona. (11) The basic design of these dams can be seen in Figure 20 [not reproduced here]. (12) The few number of these dams that were built made stability design a matter of calculation rather than experience. (13) The Hauser Lake Dam failed due to overturning.

2. The following email was sent by the head of the Employee Fitness Center on a university campus to "all employees" about a new policy (although it isn't easy to see that). The subject line was "Employee Fitness Center." Revise the text and the subject line to make the main point stand out. Hint: also revise the distribution list.

Date: January 10, 2022

The University administration has examined carefully its policies regarding use of the Employee Fitness Center (EFC). Believing that exercise is an important contributor to the health of University employees, we want to make the EFC universally available at no expense to the employee. In order to accomplish those objectives while being mindful of safety while exercising, we have decided to implement the following changes this year.

First, while we cannot increase the staffing and supervision while continuing to offer this service free of charge, we do want to make users aware that they exercise at their own risk. Each user should consider whether, in light of the fact that the EFC will be unsupervised at times, he or she would be more comfortable using a private, commercial fitness facility. If you decide to continue to use the EFC, we suggest that you do so when others are likely to be using it as well.

Next, in order to make sure that all users understand this cautionary advice, we will require that all users sign a waiver before further use of the EFC on or after February 5, 2012. With this waiver, you may continue to use the EFC at your convenience.

We will be posting signs in and around the EFC reminding all users that they exercise at their own risk. Have a healthy year and we look forward to seeing you all!

3. Select a paragraph or two from an author whose work you like. The content should be somewhat technical, not fiction, but otherwise the topic is open. Then, copy the paragraph(s) by hand, line by line. Physically writing text in this exercise has been shown to have good effects in learning about style. Then, try to write about a different topic, in this style. As you do so, think about the detailed analysis of sentence and paragraph structure you read in Chapter 10 to see how the author's style may differ from how you tend to write.

FOR COLLABORATION

If you are preparing a collaborative report with other team members in the class, create a "design specification" for its layout, use of typography and color, headings and other graphic signals for showing the relationship among parts, and its overall design persona. Create a wireframe (Chapter 4) or other sketch of a sample page or screen which can serve as a template to demonstrate your collaborative decision about design. In addition, design a cover, table of contents page, and other front and back matter as required by the audience for the report. The specification is an internal document for your team, but you may show it to others, including your instructor, for further comments. Chapter 16 provides guidelines for formal reports.

REFERENCES

AIA. (2020). AIA *framework for design excellence*. Retrieved from https://bit.ly/3tEpSnH

Davis, H. (2020). *Working cities: Architecture, place and production*. New York and London: Routledge.

Norman, D. (2004). *Emotional design: Why we love (or hate) everyday things*. New York: Basic Books.

Walter, A. (2011). *Designing for emotion*. New York: A Book Apart.

Part 4

Applying Design to Technical and Professional Communications

12 Career-Related Communications

You probably don't need to be convinced that one of the biggest tests of your communication skills will be your performance as a job applicant, in both what you write and how you conduct yourself in conversations and interviews. The design-thinking-enhanced strategies you've learned in this book will help you engage in this process with both confidence and success. This chapter provides guidelines for creating persuasive job application materials while you learn about career options, focus on particular jobs of interest, establish a professional persona, write letters of application and résumés, correspond as needed in other formats, and participate in interviews. The advice pertains as well if you are seeking an internship or applying to graduate school.

THE JOB SEARCH AS A LEARNING PROCESS

You can deploy a design-driven approach to your job search. Let empathy be a constant attitude in your search process as a reminder that employers have specific needs and you might be a solution to them. Remember that getting a job is not a one-way street. You are the embodiment of specific knowledge, skills, and experience. Employers seek candidates who can supply these attributes to grow their organization. This section teaches you how to learn about yourself and establish a professional persona to communicate your strengths.

Learning about Yourself

The job search process begins with learning about the job market requirements and resources that are available to support your search. Although it can be tempting to dive right into job posting boards or classified recruitment announcements, you should take the time to first understand your own strengths and interests before beginning to look for opportunities. Analyze your accomplishments, activities, skills, strengths, and weaknesses. Ask questions such as the following:

- What do the decisions I've made in the last year say about me?
- How do I feel about working on teams?
- Am I a risk taker or do I prefer security over surprise?
- Do I prefer on-site, remote, or a hybrid office?
- Am I willing to travel for a job?

- What's my definition of success?
- What financial obligations do I have?
- What am I most interested in or passionate about?

A number of sites offer self-analysis techniques; among these is the Clifton StrengthsFinder. The diagnostic uncovers your unique rank order of 34 themes focused on strategic thinking, relationship building, influencing, and executing. Another popular tool is the Gabriel Institute Role-Based Assessment, an online behavioral assessment for workforce hiring, coaching, and team-building. Using these analysis tools, you can get a general description of your strengths, and, in many cases, the information will help you to better match your interests and strengths with professional opportunities. You may also elect to use these descriptions as key words in your professional correspondence and social channels, including email signatures and social profile pages to highlight your strengths.

Almost all universities and colleges have a career services center that provides resources to students in preparation for their job applications. It is wise to take advantage of these services and speak with a career counselor. Some institutions have specialized career centers that are dedicated to specific majors or student populations (i.e., first-generation students, international students, or military veterans). If you are unsure which services you should approach, consult your academic advisor or a professor for suggestions.

In addition to the career centers, consult the following sources of career information to learn about occupational titles, salaries, and recommended qualifications:

- Annual reports of companies that interest you
- State and federal employment offices such as USAJobs (www.usajobs.gov) that can help you find job openings within hundreds of federal agencies and organizations
- Private for-profit recruiters
- Non-profit organizations (including adult literacy centers)
- The professional society that serves your discipline, such as the Society for Technical Communication (www.stc.org), User Experience Professionals Association (https://uxpa.org), and American Marketing Association (www.ama.org).

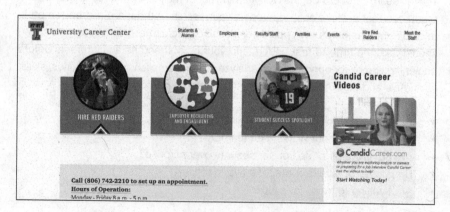

12.1
Resources Available at a University Career Center

Establishing Your Professional Persona

You may have learned from many professional development resources that advised job candidates to establish a compelling professional "brand" through online presence. It's true, recruiters are on the lookout for potential employees through online channels in the digital age. This is why you should pay attention to your digital footprints and presence on various social media during the job search process and craft a desirable persona that appeals to future employers.

Although social media platforms come and go, and their popularity is dependent on their user base, the practice of online social networking is deeply embedded in the fabric of our everyday lives. Job candidates should choose appropriate digital platforms that allow them to receive news about job opportunities, connect with businesses and organizations, share professional contents, and apply for open positions.

LinkedIn is one of such online professional networking platforms. You may create a profile with information about your educational background, work experience, interests, and skills set. Advance features include InMail messaging with recruiters, giving and receiving professional recommendations, and earning certification in specialized areas (e.g., project management, workshop leadership, digital marketing, etc.). However, crafting a persona requires more than just creating a profile using these features. Job candidates should remain engaged with regular contents and news on their channel and showcase relevant achievements to establish an outlook that entices hiring managers.

The basis of online social networking is community building. Job seekers are encouraged to use multiple platforms to establish a presence within the professional communities they aspire to be a part of. For example, *Twitter* and *Facebook* allow users to create special interest groups that serve as forums for professionals to exchange ideas and discuss issues relevant to their respective professions. Participating in these moderated communities can be a meaningful way to develop your presence and build credibility around your interest areas. Because these interest groups are typically interlinked, you can easily expand the reach of your online persona by participating in multiple groups.

Outside of these social media channels, you may also want to join professional mailing groups via *listservs* to receive timely postings about job opportunities. These email groups are usually hosted by certain institutions or individuals via internet list managers. Once you've subscribed to a listserv, you may read and post content to all subscribers of the list. To maintain a positive persona, you should always be courteous and respectful toward other list members, especially when you engage in discussions or debates around issues.

For more about online correspondence, see Chapter 18

A NOTE ON PRIVACY AND SECURITY

Keep in mind that any social media sites or online channels that support "free membership" are typically supported by third-party advertisements that allow advertisers to target their users with the data they collect through user profiles, resulting in the site's ownership of your personal data and information. You should make informed decisions about giving your personal contact and private information (e.g., interests, location, family connections) to these social media sites. Increasingly, sites also keep track of your contributions and online responses. To protect your reputation, don't get involved in inappropriate or negative conversations online.

Of course, outside of the digital realms, you should develop and maintain a professional persona in the "real" world as well. A good way to stay abreast of current events and industry trends is by attending professional get-togethers such as practitioner meetups, conferences and summits, and professional development workshops. Whether participating in person or virtually, you should consider these occasions as opportunities to learn and even share your knowledge about your professional interests (e.g., giving a short talk at a conference). Employers are often recruiting new talents at these professional events. If you do not know where to look for these events, consult your professors or career counselors for points, or check out the social networking channels mentioned above for leads. Additionally, websites such as *Meetup.com* and *Eventbrite.com* offer services to organize these events, and so it is a good idea to subscribe to them for notifications about upcoming opportunities.

ORGANIZING AND PRESENTING JOB MATERIALS

Getting yourself seen is only the first step. To begin the job application process, you need to create an effective organizing system, understand job descriptions, gather recommendation letters, and, of course, compose your résumé and application letter. This section covers these important steps in the job application process.

Creating an Organizing System

Whether you are searching for a job or applying to graduate school (which requires similar professional documents), the process is a job in itself. It demands careful planning and accurate record keeping of:

- Job leads and requirements
- Graduate program descriptions
- Sample work and portfolio
- Letters sent and received
- Recommendations/references
- Interview appointments
- Follow-up notes.

Materials such as your writing samples, design work, and portfolio take time to curate, and so it is a good idea to start working on them before you hit the job market. During your time in the degree program, identify assignments or projects that you performed well in and save them on a drive where you can access them later when building a professional portfolio.

Use cloud-based storage or a file management system to save and organize your materials. Most email servers (e.g., Google Gmail, Microsoft Outlook, Apple iCloud) come with a cloud drive with decent storage capacity where you can upload your job materials. Organizing your materials this way has several benefits. First, you do not need to worry about losing them from a local hard drive (i.e., your laptop computer) or an unfortunate file crash. Second, you can manage others' access to your materials by controlling the sharing options on these drives. Relatedly, you can generate web links (URLs) to your materials from these drives and share them easily on your online social profiles and with interested recruiters. Third, these cloud storage tools also let you gather feedback from your professors or counselors through the collaborative authoring and commenting features in the system. You can allow viewers to add comments or suggest changes on your materials that you may use or discard as you refine your materials.

Finding and Understanding Job Descriptions

A job description in an announcement of an opening provides details about an open position, explaining its function, expected scope of work, and the qualifications required to fulfill the work. Most job descriptions include information about the application process and the personnel that applicants could contact regarding the open position (typically someone from human resources).

Here are some popular websites on which to locate internship opportunities and job postings:

- *Indeed* (www.indeed.com/), the largest employment-related search engine
- *SimplyHired* (www.simplyhired.com/), an online employment and recruitment advertising network
- *ZipRecruiter* (www.ziprecruiter.com/), an online job board
- *Internships.com* (www.internships.com/), a database specific to college-level internships.

When reading job advertisements and position descriptions, remember that employers have specific needs and are looking for talents to address those needs. Applying an empathy mindset, you should look for key words in the description that indicate these needs and use them in your résumé and reference letters. This lets the employer know that you care about their specific hiring goals and that you are prepared to help meet those goals.

FINDING SALARY INFORMATION

Not all job advertisements include specific salary information for a position. There are a few reasons for this, one being the employer's hope to avoid deterring potential candidates owing to unattractive compensations. However, you as the job seeker can research the salary range based on the position type on job listings and companies review websites such as Glassdoor.com and Indeed.com. You may also ask friends or family members, as well as the general communities on your special interest social media groups, about the general offers for a common position.

Gathering References or Recommendation Letters

Develop a list of people who can testify about your abilities, accomplishments, and potential as a graduate student or professional. In letters of reference (sometimes called letters of recommendation), recommenders comment on what you have done and, thus, what you can do. Because their own credibility enhances the force of their recommendation, request letters from people whose word will be taken seriously:

- Your adviser or major professor
- A supervisor who can describe your work habits
- An adviser to a campus organization or sports team in which you have been active
- People you know who also know the university, organization, or company to which you are applying.

In asking for a letter of recommendation, keep the following guidelines in mind:

- *Ask first:* Never mention a recommender to a potential employer without checking first with the person you've named. The step is both courteous and self-serving. Some job candidates ask directly if the person would be able to write a strongly positive recommendation.
- *Brief the recommender.* Convincing letters are concrete and detailed. Talk with your recommenders about your intentions and provide them with a résumé, a statement of your goals, and some background on the company or graduate program to which you are applying.
- *Do your homework:* Complete sections of recommendation forms that require your own information or signature. Do whatever you can to make the logistics of filling out the forms easy for the recommender.

Composing Your Résumé

A résumé has a simple purpose: to list what you have done as a predictor of what you can do. Although many hiring platforms such as Indeed.com, LinkedIn, Monster.com, and Interfolio (for academic jobs) offer electronic services that gather and organize your personal information, a résumé is still a necessary arti-fact that most employers require in the job application packet. To create a résumé, start by making a list of experiences that you have had—education and training, jobs, internships, research, projects, volunteer work, leadership roles, student organizations, etc. Think about what you have contributed to these, the skills you used and developed, and significant achievements or goals you have met from these experiences. Then, begin to compose your résumé by organizing these into sections.

Although readers expect certain units of information, content and design are not universally dictated and can differ between print and online formats (Smith Diaz, 2013). Two approaches are common: chronological and functional.

Chronological Résumé

In a chronological résumé, you list your education and experience by order of time. Figure 12.2 provides an example of this type. Most students find this approach the most useful.

12.2
Chronological
Résumé

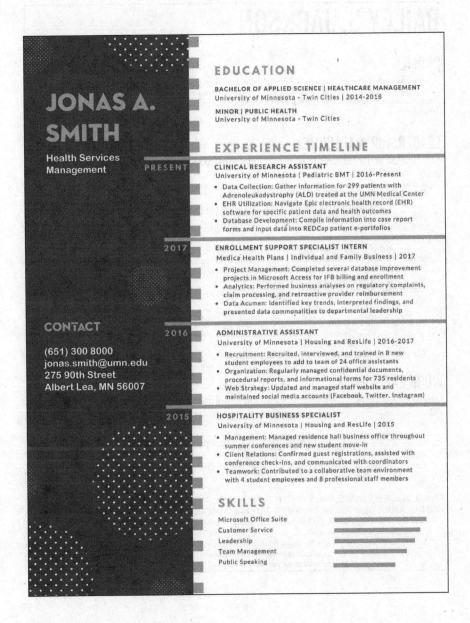

JONAS A. SMITH

Health Services Management

CONTACT

(651) 300 8000
jonas.smith@umn.edu
275 90th Street
Albert Lea, MN 56007

EDUCATION

BACHELOR OF APPLIED SCIENCE | HEALTHCARE MANAGEMENT
University of Minnesota - Twin Cities | 2014-2018

MINOR | PUBLIC HEALTH
University of Minnesota - Twin Cities

EXPERIENCE TIMELINE

PRESENT
CLINICAL RESEARCH ASSISTANT
University of Minnesota | Pediatric BMT | 2016-Present
- Data Collection: Gather information for 299 patients with Adrenoleukodystrophy (ALD) treated at the UMN Medical Center
- EHR Utilization: Navigate Epic electronic health record (EHR) software for specific patient data and health outcomes
- Database Development: Compile information into case report forms and input data into REDCap patient e-portfolios

2017
ENROLLMENT SUPPORT SPECIALIST INTERN
Medica Health Plans | Individual and Family Business | 2017
- Project Management: Completed several database improvement projects in Microsoft Access for IFB billing and enrollment
- Analytics: Performed business analyses on regulatory complaints, claim processing, and retroactive provider reimbursement
- Data Acumen: Identified key trends, interpreted findings, and presented data commonalities to departmental leadership

2016
ADMINISTRATIVE ASSISTANT
University of Minnesota | Housing and ResLife | 2016-2017
- Recruitment: Recruited, interviewed, and trained in 8 new student employees to add to team of 24 office assistants
- Organization: Regularly managed confidential documents, procedural reports, and informational forms for 735 residents
- Web Strategy: Updated and managed staff website and maintained social media accounts (Facebook, Twitter, Instagram)

2015
HOSPITALITY BUSINESS SPECIALIST
University of Minnesota | Housing and ResLife | 2015
- Management: Managed residence hall business office throughout summer conferences and new student move-in
- Client Relations: Confirmed guest registrations, assisted with conference check-ins, and communicated with coordinators
- Teamwork: Contributed to a collaborative team environment with 4 student employees and 8 professional staff members

SKILLS

Microsoft Office Suite
Customer Service
Leadership
Team Management
Public Speaking

Functional Résumé

In a functional résumé, you categorize information about education and experience under headings that indicate your skills and achievements (Figure 12.3). Those categories usually reflect requirements stated in a job description—for example, "team-building skills," "management experience," "information technology." Use this approach when your experience is scattered in type and amount of time, you are shifting careers, or when you are seeking a narrowly defined job, internship, or training program for whose specific requirements you need to show a match.

BAILEY S. JACKSON

PROFILE

I am a marketing and retail merchandising student seeking a full-time summer internship position in the field of social media and marketing communications, where I can apply my knowledge and skills for continuous improvement and development.

LEADERSHIP & ACTIVITIES

DIRECTOR OF ALUMNAE RELATIONS   **SEPT 2016-PRESENT**
DELTA GAMMA FRATERNITY, LAMBDA CHAPTER
- Organizes networking events with Delta Gamma alumnae and current collegian members.
- Maintains relations between Delta Gamma alumnae and the Lambda chapter In charge of Lambda chapter LinkedIn page
- Engages in ongoing philanthropic events for the organization's Service for Sight Foundation.
- Gains philanthropic experience in participating and volunteering at chapter philanthropy events such as Anchor Bowl and Golden Anchor Ball Participates in ongoing continuing education in the areas of leadership, goal setting, and mentoring at chapter meetings.
- Understands and adheres to the ideals, policies, and values of Delta Gamma Fraternity.

SALES & LEADERSHIP CLUB   **JAN 2018-PRESENT**
- Attends seminars, networking events, and organization meetings Practices sales strategies and skills philanthropic.
- Competes in mock sales competitions involving local companies.

WORK EXPERIENCE

SUMMER CARETAKER   **JUNE 2018-PRESENT**
KELLEY BURNETT   612-770-8000 / KBURNETT@AMFAM.COM
- Safely transports children to and from school, medical appointments, and extracurricular activities.
- Organizes activities that enhance children's physical, social, and emotional wellbeing.
- Built relationships with children to teach them to be understanding and paient with others 50 hours per week.

FRINGE WOMEN'S ACCESSORIES   **JAN 2014-JAN 2018**
KRISTIN ZAZULA   612-300-3000 / KRISTIN.ZAZULA@COMCAST.NET
- In charge of opening and closing the store, handling merchandise transactions, and taking inventory.
- Involved in all aspects of fashion merchandising - help select new seasonal merchandise, create window displays, personal styling for customers including brides, plans and orchestrates in-store fundraisers.

CONTACT

(952) 300-8000
bailey.jackson@umn.edu
5990 Lane View
Edina, MN 55400

EDUCATION

BS, Business Marketing
Minor in Retail Management
University of Minnesota

SKILLS

- Strong written and oral communication
- Social media strategy
- Ability to anticipate customer needs
- Self motivation
- Proficient in MS Office
- Ability to work under pressure
- Leadership
- Decision making
- Time management
- Great eye for design

12.3
Functional Résumé

Units of Information on Your Résumé

Whether you are composing a chronological or a functional résumé, include the following units of information to organize your résumé.

Heading

Standard information in the heading allows the potential employer to file your résumé and to get in touch with you:

- Your name
- Mailing address
- Phone number
- Email address
- Personal website/portfolio link.

Design the heading to attract attention and set your style. At either the top or the bottom of the résumé, you might indicate the date on which you last updated the résumé to show its currency to both the potential employer—and you. The date line will remind you to send the current version of a document you may revise many times.

Key Words

Whether you use a chronological or a functional approach, thinking about categories of skills will help you develop a list of key words about yourself. Such terms are increasingly common on résumés, especially ones that will be scanned or read online and thus entered into a database that can be searched to fill job openings. Those key words should match the terms an employer uses to describe its search. Here are some examples of key words:

- Information systems/technology
- User/customer support
- Self-starter
- Ability to learn quickly
- Handle complex conceptual problems
- Effective team player
- Mature, experienced sales executive
- Strategic planning
- Detailed implementation
- Proven track record in [name of method or practice]
- Key areas of responsibility:

    o   Managing support services
    o   Reviewing and enhancing systems and procedures
    o   Monitoring budgets and costs.

- Challenging role.

Objective

Résumés often include a statement of objective, which many recruiters say they like to see. In creating such a statement, be concrete: note what you seek to do, not what you want to be.

Objective: to serve as a support person or aide on a physical therapy team.

Create different statements, if necessary, for different jobs. Avoid being wordy in stating the objective. If you can't define your objective in concrete terms, omit the statement.

Summary

Include a brief statement summarizing your skills and achievements if that helps you consolidate extensive amounts of experience, especially if you are changing careers:

Experience as a teacher of secondary mathematics for over 15 years. Completing a degree in accounting (career change) with the ultimate goal of qualifying as a certified public accountant.

A summary statement is also useful on a résumé that will enter a digital database.

Education

Be thorough in describing your education. As Figures 12.2 and 12.3 show, list colleges attended, beginning with the current one, and degrees expected or attained. You may also include high school if you feel identifying the school will comment strongly on your own credentials. However, generally speaking, high school experience is assumed, not expressed.

Other items you may include:

- Brief list of important courses
- Grade point average (in the major or in general)
- Scholarship awards (not including dollar amounts)
- Special honors (such as dean's list or honorary societies)
- Digital competencies (computer skills beyond common software)
- Language skills (speaking, listening, and writing)
- Undergraduate research experience
- Peer tutoring or teaching experience (in peer training, workshop facilitation, or teaching assistant roles).

Be imaginative but honest. Use this information to differentiate yourself from the pack. Indicate activities and knowledge, not just an ability to show up. For example, rather than noting that you have had six semesters of French courses, indicate fluency in French, if indeed that's what you achieved. When you state that, of course, be prepared to respond *en français* if an interviewer greets you with *Bonjour*. List courses by their substantive titles ("Thermodynamics") rather than the university's internal code ("Eng 421"). Avoid acronyms that a potential reader might not know. A civil engineering student applying for jobs in civil engineering can feel comfortable listing the EIT (Engineer in Training) test but should explain the acronym for another reader.

Experience

Describe your work experience in detail, using either the chronological or functional approach. Be specific about the jobs you have held, including your title, the name of the employer, and your responsibilities. Assess the skills you demonstrated—for example, writing reports, preparing financial statements, operating equipment, or monitoring processes:

> Davidson Laboratories Salt Lake City, Utah
> June to September 2020
> Job title: Writing Intern
> Scope of Responsibilities:
>
> - Designed five online documents for the laboratory's website
> - Researched medical topics and published two articles for a general audience
> - Worked closely with the medical illustrator
> - Delivered final product formatted in HTML.

Note any increase in responsibility that shows success in your role. Where possible, quantify the results of your work—for example, an increase in sales or a decrease in the amount of organizational paperwork attributable to a software system you designed or installed. Describe any management or supervisory role you played. Note both independent work and teamwork. If you participated in an internship as part of your academic program, you may categorize that work as either experience or education, depending, in part, on which unit of information needs supporting entries. Even if the content of the jobs you've held doesn't resemble what you want to do in your career, your ability to work effectively in an organizational context speaks well of you to a future employer. In describing your experience, note any special skills you possess or levels of certification you have achieved. For example, if you are interested in a career in golf course management, note your golfing handicap training.

Personal Background

Somewhere in the résumé you may want to place information about your interests and self-development activities (including sports, special hobbies, and even travels). In an increasingly global economy with online job searches that span nations and continents, indicating you are willing to relocate can also be a selling point. Weave such information into the education and experience segments or set up a catchall category at the end of the résumé. You don't need to include such details as your birth date, marital status, ethnicity, and religious preference. US affirmative action guidelines prevent employers from asking for this information. But you may find that employers in other countries expect to see such facts on a résumé.

References

To save space, most people simply indicate on their résumé that letters of recommendation are available upon request. But an impressive list of recommenders may have persuasive value. In addition, conventions differ from discipline to discipline. If the job announcement requests it, include the names of recommenders in your letter of application. Otherwise, select particular recommenders to match particular jobs; indicate their names in a cover letter or note the availability of a dossier at your school.

Designing Your Résumé

How you package the units on your résumé depends on the conventions of your discipline as well as your own design preferences. As you design your résumé, look at the résumés of others who were successful at seeking the kind of job you seek. Emphasize the units that best showcase your abilities. Start, for example, with education if that's your strong suit, or with experience, if that's more persuasive.

For more about composing visuals, see Chapter 9

The résumé stands for you—on paper or on a screen. It should convey an appropriate persona and it should also be easy to read, two purposes that are sometimes in conflict. Employers give résumés a notoriously short amount of time to make their point, so appropriate emphasis on key credentials and readability is essential. You want to make sure your assets as a candidate are obvious, so spend time to achieve an attractive and efficient design. Spend time, too, to ensure that your text is letter perfect. A spelling error or simple error in grammar provides an easy way for a reader to eliminate your résumé; if you are not careful with a résumé, the thinking goes you will not be a careful and precise technical professional on the job.

Print

Figures 12.2 and 12.3 provide good models of printed résumés that take advantage of the options available in desktop publishing software. Keep such résumés short—generally one page, sometimes two. In your attempt at brevity, don't be lured, however, below a 10-point type size for major text. If the type is too small, it becomes difficult to read, and you've lost the advantage of brevity. Condense the text instead. Use graphics and layout to enhance a personal statement. Create an original résumé with a high-quality printer and reproduce it on white or off-white bond paper.

CONTINUOUS IMPROVEMENT

A résumé is not a static document. It grows with your experience and professional development. Apply the design-thinking mindset of ideation and iteration to continuously improve your résumé in terms of content as well as design.

Setting Up an Online Portfolio

If you are seeking a job that requires digital skills, having an online portfolio helps convince a prospective employer that you have those skills and are comfortable with digital media (Watson, 2019). A portfolio is a collection of materials that exemplifies your skills, qualifications, and experiences. It may include your résumé, writing samples, design projects, training certifications, and narratives about your career goals and work philosophy. An online portfolio allows you to showcase these materials digitally—leveraging the affordance of multimedia presentations such as videos and audios—and make them accessible to employers.

To start, you need to select a content management system (CMS) that gives you a template to begin this process. CMSs such as *Wordpress* and *Google Sites* are relatively user-friendly: they do not require coding skills to use the platforms. You may also elect to use WYSIWYG (what-you-see-is-what-you-get) tools such as *Wix* and *Squarespace* to create your own layout with drag-and-drop design features. Other services such as *Adobe Portfolio*, *Behance*, and *Digication* give users readily available gallery displays for visually driven projects (e.g., graphic design, user-interface design, photography).

Regardless of the platform you choose, you should be intentional about the organization (content and layout) of your portfolio, and how you narrate your writing or design process. Ultimately, an employer reading your portfolio aims to understand how you define problems and ideate solutions. By the way of describing selected issues and your approach to addressing them, you can demonstrate critical and design thinking throughout your portfolio.

When it comes to the amount of information, moderation is key. Because the range of available tools may be wide, you'll need to identify carefully what is appropriate content and avoid a tendency to overwhelm your readers online. To encourage employers to visit your website, include the URL to your website on your print résumé.

Writing an Application Letter

The résumé provides the facts backing your assertion that you are the best person for a particular job. In a letter you send with the résumé, called a *letter of application* or a *cover letter*, you argue the truth of that assertion. You also establish your character and persona. As an example of the importance placed on character in hiring decisions, tradition in France dictates that letters be handwritten, even by those who are digitally literate.

Read job descriptions (for examples, see the boxes) carefully when you write your application letter to make sure you have responded to each requested credential. Keep in mind the who, why, and what approach to create an effective letter.

SAMPLE JOB DESCRIPTION

Tyler Technologies is looking for a talented, passionate Software Engineer to enhance and maintain our Law Enforcement and Fire Records, and Corrections

Management Systems. As a Software Engineer, you will work in a fast-paced, results-driven environment with a highly skilled and dedicated team to implement solutions as well as remediate client-reported issues. Ideal candidates will enjoy solving complex technical problems and thrive in a highly collaborative environment.

Incode Public Safety offers mobile communications, comprehensive records management, Geographic Information Systems (GIS) to track mobile unit activity, and business/residential alarm system registration. This public safety solution conforms to Uniform Crime Reporting (UCR) standards and integrates with the National Crime Information Center (NCIC). Incode Public Safety is a public safety solution used around the country by police, tribal police, campus police, transit authority police, sheriff, and postal police departments as well as fire and rescue units.

Responsibilities

- Participate in the entire lifecycle of analysis, design, coding, testing, implementation and support
- Develop new features as well as maintain legacy code within the product suite
- Ensure on-going success of projects by designing high-quality technical solutions
- Collaborate with other software developers, business analysts and software architects to solve complex technical problems
- Participate in troubleshooting of production issues
- Lead and mentor Junior Developers and/or Interns

Qualifications

- 2+ years of experience in designing and programming applications
- Strong knowledge of computer science fundamentals in data structures, algorithms, complexity analysis, and databases
- Exceptional software design, problem solving and object-oriented coding skills
- Skills and experience with C#, SQL, VB Script, XML/XSD/XSLT, DHTML, Visual Basic 6
- Proficient with tooling to enable SDLC (TFS, Visual Studio, etc.)
- Ability to excel in an Agile based team with a strong focus on collaboration and teamwork
- Strong knowledge of design and code patterns, specifically toward .NET
- Ability to understand and follow existing architectural patterns
- Positive outlook and willingness to learn and accept feedback from others
- Strong communication skills especially around technical team interaction
- Familiarity with Test-Driven Development (TDD) and test automation frameworks is helpful

- Demonstrated experience developing enterprise business applications is strongly preferred
- Bachelor's Degree in Computer Science or related technical field, or equivalent work experience

RESEARCH ASSISTANTSHIP DESCRIPTION

Undergraduate Research Assistant Hourly Position
Application due **August 24, 2021**

The TTU Technical Communication Program seeks one (1) Undergraduate Research Assistant (RA) to work in collaboration with the grant-funded project, "Understanding the Design, Delivery, and Impact of Multimodal Social Advocacy Projects." The student hired into this position will work closely with collaborators from other institutions.

Workload
- Project coordination (10%)
- Data collection (30%)
- Analysis (60%)

Duties
The RA will assist with the following tasks:

- Reviewing and summarizing relevant literature sources
- Managing and preparing materials for submission to granting agency
- Using an online transcription service to transcribe interview sessions into data scripts
- Analyzing participant interviews
- Preparing, maintaining, and updating materials on cloud drive and project website
- Preparing and reviewing articles, reports, and presentations

Required Qualifications
- Candidate must be a registered undergraduate student at Texas Tech University, Fall 2020, and be registered full-time and be making adequate progress toward degree completion
- Familiarity with networked communication technologies including emails, scheduling tools, and video conferencing
- Ability to work individually and collaboratively as a team member
- Ability to liaison with faculty, instructors, and community partners
- Strong communication skills

Appointment and Pay Rate

- This is an hourly appointment (up to 20 hours/week) for 10 weeks, Fall Semester 2020, starting 9/1/2020 and ending 11/6/2020.
- The hourly rate is $12.00 with a maximum salary of $2400 (200 hours total).

Application Process

Deadline for applications is **August 24, 2021.** Please send a résumé and cover letter via email to **Dr. Jason Tham** (jason.tham@ttu.edu); refer to any relevant work you have done and an explanation of why you are a good fit for the project. Interviews will be conducted via use of Zoom. If hired, you must complete an I-9 form and provide original document(s) verifying your eligibility to work in the US by the start date of your appointment (9/1/2020).

Texas Tech University is an Equal Employment Opportunity/Affirmative Action employer. The university does not discriminate on the basis of an applicant's race, ethnicity, color, religion, sex, sexual orientation, gender identity, national origin, age, disability, genetic information or status as a protected veteran.

Who

First, identify yourself as a responder to a specific job opening, and then match your qualifications to those requested. Although the letter is about you, avoid opening every paragraph with "I." In particular, avoid a flat first sentence such as, "I will graduate from the University of the West with a degree in civil engineering." Instead, interpret yourself in the reader's terms. Consider these more reader-oriented openings:

- *Reference to an advertisement:* I am writing in response to your advertisement for an auditor in the 16 May edition of the *Wall Street Journal.* I hope you will agree that my education and experience qualify me for your position.
- *Reference to an individual:* Professor Peter Prince of our Department of Chemical Engineering suggested that I apply to you for the opening in your environmental engineering project office.
- *Reference to past experience with the company:* Having spent the last four summers as a playground director with your department, I would like to apply now for a permanent position as director of one of your recreation centers.

Why

Next, match yourself to the potential employer and, thus, merge *who* and *why*. Address your letter to a specific person whenever possible. If you hear about a job from a friend, former employer, or professor, be sure to ask about the name of the person to write to. Then emphasize your interest in and understanding of the reader and his or her organization. Tailor the facts in the résumé to the requirements of the job you are applying for. Show the reader what you have done and can do. Avoid mere generalizations. A statement such as "I am conscientious and hardworking" carries no conviction, but specifics do: "I have maintained a grade point average of 3.8 (on a 4.0 scale) while working 20 hours a week as an assistant in the chemical engineering laboratory."

- *General*: I get along great with people.
- *Specific*: I am president of my senior class and have held two offices in my sorority: secretary and treasurer. Currently, I serve on the program committee for Engineers' Day, which entails working with 24 other students and 5 faculty members.

- *General*: I know a lot about designing children's furniture.
- *Specific*: My interest in designing children's furniture was triggered by my visit to the Furniture Mart in Chicago three years ago, where I saw several pieces designed by your company. I then switched my major from architecture to design. In my design classes, I focus any open assignments on issues in designing for children. I wrote one report, for example, on school playgrounds in Denmark. My design for a playbox was featured in our student exhibit last year.

What

Finally, request an interview, the goal of your letter, and refer to your enclosed résumé:

> The prospect of working for DelMart is most appealing. If my enclosed résumé interests you, I would be happy to talk with you further at your convenience.

If you live some distance from the potential employer, mention that you would be willing to travel there for an interview (if you are) or would be available for a phone interview, and indicate appropriate times.

Follow Up after Your Application

If you do not receive updates after about 2 weeks of submitting your résumé and application letter, it is appropriate to check on the status of your application with a follow-up letter or email. This gesture encourages the hiring manager to attend to your application materials and give you a timely response. Be brief and clear in your follow-up message:

Dear Ms. Johnson:

On January 5, I submitted a cover letter and résumé for the UX designer position advertised on Social Company's website. As of January 20, I have not received further communication from your office. I want to confirm that you received my application and restate my interest in this position.

I remain enthusiastic about the opportunity to work at Social Company, and I believe my design education and internship experience have well prepared me for this role.

I look forward to discussing my qualifications with you further. Thank you for your consideration.

Sincerely,
Your Name
yourname@email.com
222-555-8888

For more about professional letters, see Chapter 18

**12.4
Letter Responding to Research Assistant Position**

Source: Courtesy of Liane Vásquez-Weber.

Use proper letter format to create your cover letter

Lubbock, TX
(830) 000-0000
l-v-w@ttu.edu

August 18, 2021

Dr. Jason Tham
Department of English
Texas Tech University
P.O. Box 43091
Lubbock, TX 79409

Dear Dr. Jason Tham:

Reference how you learned about the position

I am applying for the Undergraduate Research Assistant Hourly Position for the grant-funded project, "Understanding the Design, Delivery, and Impact of Multimodal Social Advocacy Projects." Ms. Akshata Balghare shared the job description with me. As a writer majoring in Computer Science I am thrilled to apply for this position!

Mention specific experiences that match the requirements of the position

As team-lead for my Technical Writing final project, I had to adapt and be flexible to work with every team member and keep all of us on the same page, especially as we transitioned online mid-semester. My default approach to projects is to discover the creative ways to accomplish assigned tasks. For example, I suggested my team and I create technical documents for a real-life volunteer organization I knew, instead of an example case study. Our client was greatly satisfied with the work we accomplished. It is due to my creative approach to problems that I also am a part of an ongoing research project at the intersection of the fields of computer science, biology, and neuroscience.

I have experience with reviewing and summarizing literature from the above research project, as well as from a biology course I completed. I am analytical by nature, able to name and address the pieces of a problem and steps toward a goal. This strength helped me in my service-learning course in Educational Processes as I planned and executed an original online workshop for my target audience. My verbal and written communication skills are exceptional, as I lead with confidence, professionalism, and warmth. These communication skills qualified me to become the Social Chair for a student organization.

Close the letter with a request for interview and refer to enclosed resume

I am uniquely poised for this position, and I look forward to discussing my qualifications further in an interview. Included also is my resume with relevant experience and work. Thank you for your time and consideration. I look forward to hearing from you.

Gratefully,
Liane VW
Liane Vásquez-Weber

Enclosure: Resume

INTERVIEWS

Once hiring managers have reviewed the application materials, successful candidates will be contacted for an interview. You may expect a notification for an interview within 2–4 weeks of submitting your application. Typically, this notification will either be emailed to you or made through a phone call. To prepare for the latter, be sure to set up your phone voicemail greeting (with information about when to call back, should you miss the call). In this section, you will learn about tips and strategies to prepare for a phone, video, or in-person interview, and how to follow up with thank-you notes and letters of acceptance or refusal.

Preparing for Your Interviews

In preparation for your interview, you should carefully review the job description and note your key qualifications for the open position. Perform research on the company's history and culture, leadership and administrative structure (key personnel and teams), current projects, and recent achievements. Envision what you can offer to the company, both practically now and what you might aspire to. For example, you may consider how you can serve in the role that is advertised and how you can provide even greater value to the company through future undertakings. These considerations may come in handy when you are asked to suggest improvements to the company—a common interview question. Additionally, you should be prepared to respond to the following questions with ease and confidence:

- Tell us about yourself and why you are interested in this position.
- What are your greatest strengths?
- What do you consider to be your weaknesses?
- What are your proudest accomplishments?
- Tell us about a time you demonstrated leadership skills.
- Tell us about a time when you experienced a challenge and how you overcame it.
- How would your peers describe your work ethic?
- How do you deal with ambiguous or difficult situations?
- Where do you see yourself in 5 years?
- What questions do you have for us?

ALWAYS BE TESTING

There are many strategies to prepare for an interview; the most important yet is to test your materials and readiness. Using the design-thinking mindset of testing, you may rehearse the interview session with a friend or family member. This would allow you to practice the actual words you plan to use in your answers, the time you take to respond, and the notes you may need to aid your responses.

Interviewing by Phone

Some organizations may conduct multiple interviews to screen candidates for a position. The first of these interviews is usually conducted via phone. To prepare for a phone interview, you should practice answering questions over the phone to help you recognize any verbal tics, enunciation issues, or speaking speed. You may ask a friend or family member to conduct a mock phone interview and record it so you can replay it for review.

Your voice represents your persona in a phone interview. Try reducing verbal buffers such as "ums," "uhs," and "okays" to avoid distracting the interviewer. Practice the pacing of your responses to make your answers comprehensible and make room for the interviewer to ask questions. When practicing your interview, pay attention to your listening and speaking habits. Do not interrupt the interviewer when they are asking questions and avoid rambling when answering questions.

TIPS FOR PHONE INTERVIEW ENVIRONMENT

Take your call in a room or space free of background noise or other distractions. Let your roommate or family know when you're expecting a phone call so they can leave you to answer the call in peace. Have some writing and notetaking utensils at your disposal so you can jot down thoughts or questions during the interview, and find time to ask them during the interview.

Interviewing over Video Conferencing

Video interviews are becoming commonplace owing to the popularity of video conferencing technologies such as Skype and Zoom. Video interviews are a hybrid of phone and in-person interviews. You should practice the strategies listed for phone interviews while keeping in mind that video interviews include the visual elements of your persona, in addition to audios.

The process for a video interview may be as follows:

1. A hiring manager will let you know of the interview date and time (be sure to confirm or convert time zones if your interview is not local).
2. You will be sent a link to "join" a video conference room.
3. Once you are in the virtual conference room, you will see the interviewer(s) and may be asked to be recorded using the software feature.
4. During the interview, you are expected to use both your camera and microphone to present yourself. You may also use the chatroom and screen-sharing features (depending on software) to showcase examples of your work.
5. When you are done, you can leave the interview by exiting the virtual conference room.

Dress professionally for your video interview. Avoid "busy" patterns on your outfit (including stripes) as they may cause visual effects or movements on video that can be distracting to your interviewer. Plain, neutral colors are known to appear best in videos.

Use a laptop or desktop computer, rather than your phone or tablet, for greater stability and visibility of system features. Walk through your setup, including the audio settings (speaker and microphone volume, feedback), camera (angle, focus), and lighting. A good setup should place you in the center of your video, with soft lighting so you appear personable. Place your camera a little above your eyeline so you appear upright, not staring up or downward at your interviewer.

If you use an external/extended monitor (as in Figure 12.5), you may pull up notes that can aid you during the video interview. However, be sure not to be distracted by the secondary screen and remind yourself to stay focused on the interviewer at all times.

TIPS FOR VIDEO INTERVIEW ENVIRONMENT

Similar to a phone interview, be sure to conduct your video interview in a room or space free of distractions or noise. Pay attention to the background in the video to make sure there is nothing distracting. A plain wall or bookshelf is commonly acceptable. If there are elements you can't control in the backdrop, consider getting a simple folding screen to set behind you to block out those elements.

12.5
Simple Setup for Video Interview
For a better video interview experience, use an external ring light, built-in camera (laptop computer), and external monitor.

As with any other interview preparation, you should find someone to practice the video interview process with. Consider how you look when you're listening to interview questions, and how your hand gestures appear in the video during practice. If you realize you need accommodations for video interviews, owing to visual, hearing, or other constraints, be sure to communicate that with your interviewer before the interview. Most video conferencing software has captioning and other accommodation features that can help alleviate these constraints.

For more about synchronous presentations via video conferencing, see Chapter 19

Interviewing in Person

For an in-person interview, prepare to engage your interviewer with your confidence and readiness to hold unmediated conversations. The difference between an in-person interview and phone or video interview is that your performance is evaluated in real time with relation to your physical behaviors and interpersonal skills.

As with phone and video interview preparations, you should still be familiar with the job requirements and anticipate questions about your qualifications. However, focus your practice on your demeanors and interactions with an interviewer. Pay attention to nonverbal cues and establish rapport with your body language. Practice giving a firm and confident handshake. Show your interest in the interviewer by engaging eye contact and listening attentively during the interview.

RECONSIDERING THE HANDSHAKE

The global health pandemic that began in 2019 has called into question the conventional handshake. While it has always been a cultural concern (i.e., some cultures avoid physical contact, including handshakes, even in professional settings), the COVID-19 virus has heightened new worries about bacteria transmission through handshaking. Whenever meeting with a potential employer, always research first about their handshake preference: you can ask a hiring manager about this preference or speak with a current employee. Alternatives to handshakes include head nods, a signed "thank you," and light bow. At any rate, demonstrate respect through eye contact or other verbal expressions.

Be prepared to present any documentation or physical portfolio during your interview. You may consider using a briefcase to carry these materials to your interview. If you have something that you would like to show on a screen, make sure the materials (e.g., videos, presentation slides) are preloaded on your computer or tablet before the interview so you do not need to rely on the internet to download them.

Follow Up after Your Interview

Thank You Notes

After the interview, send a brief note to the interviewer that expresses your appreciation, reminds the interviewer about yourself during that critical time when a decision is being made, and reiterates your interest in the position. Be brief but cogent:

> I enjoyed talking with you yesterday at your Smeaton plant. Touring your extensive facility and meeting potential coworkers only increased my interest in joining Peterborough Technologies. I was particularly impressed by your emphasis on quality control and team management.
>
> Thank you for taking the time to arrange such an informative visit and especially for your personal attention to all my questions. I look forward to talking with you again.

Accepting an Offer

When you are offered a job that you had set your sights on, celebrate, and then write a letter of acceptance. Briefly express your pleasure in being offered the job, show that you are eager to work for the company, and confirm the starting date and any other conditions of employment that require such confirmation:

> I am delighted to accept your offer to join Peterborough Technologies as a software designer.
>
> As you suggested. I'll plan to be in Smeaton on Friday, June 15, so that I can take up my duties the following Monday.
>
> Enclosed are the completed health survey and the signed statement concerning proprietary information.
>
> If you need any more information from me, or if you have further information for me, please write or call. I'll be at this address until June 13, when I'll begin my relocation. I'm looking forward to working with you.

Declining an Offer

When you decide not to accept a job offer, write a brief letter conveying and explaining that decision:

> I very much appreciate your offer of a position as a software designer. For the past several days, I have given the matter serious consideration, but I have finally decided that I cannot accept your offer, principally because I am not able to move to Smeaton at this time.
>
> Thank you for taking the time to meet with me and guide me around the plant. Peterborough Technologies is an interesting company, and I regret that I can't accept your offer.

Responding to a Refusal

Even if your interview does not result in a job offer (many don't), consider writing one more letter if you are still interested in the company:

> In response to your letter of 15 April, I, too, am sorry that you do not have a position for me at this time. I enjoyed talking with you and learning more about your company. Peterborough Technologies remains very attractive to me and I hope you'll keep me in mind for any future openings.

A polite letter that reiterates your interest lets you have the final word—and may also help you if a position does open up in the company.

This chapter prepares you for a design-thinking-enhanced job search. Apply the preparation strategies and guidelines to writing traditional application materials, such as résumés and application letters. These strategies are transferable to internship and graduate school applications as well. The guidelines in this chapter will also help you address the communication situations that follow your submission of an application, including interviews and correspondence.

CHECKLIST: CAREER-RELATED COMMUNICATIONS

1. Use institutional and online resources to
   - Learn about yourself
   - Learn about different career opportunities
   - See job postings
   - Establish your professional presence

2. Create an effective organizing system to
   - Keep tabs on job leads
   - Curate sample work or portfolio
   - Collect recommendation letters
   - Schedule interviews

3. Design your résumé to
   - Show what you have done as a predictor of what you can do
   - Demonstrate your persona
   - Showcase your education/training and expertise through a chronological or functional structure

4. Tailor your résumé to a persuasive argument in a letter of application
   - State why you are writing to the addressed person
   - Apply for the job

Provide evidence that you possess—or exceed—the required qualifications

Close by requesting an interview

5. Follow up with

Letters or emails to inquire about your application status

Interviews: phone, virtual/video, or in-person

Thank-you notes

Acceptance or rejection of offers

EXERCISES

1. Interview a personnel manager at your college or a local company about what they look for in a letter of application and a résumé. Here are some questions to include:

    *   Should applicants include a profile picture on their résumé?
    *   Do you require information about courses taken and grade point average?
    *   Does the résumé have to be only one page?
    *   Should the applicants include a statement of objectives on their résumé?
    *   What should the applicants include if they don't have a lot of workplace experience?
    *   How important is the design of the résumé and application letter? Do you prefer plain or visually rich résumés?

2. Comment on the following letter applying for an accounting position. The job description indicated the need for "3+ years of accounting experience, preferably in cost accounting, and demonstrated accounting software experience."

    Your advertisement on LinkedIn for a cost accountant caught my eye this weekend. I feel that my qualifications would be a very good match for your needs in this position. I will complete requirements for my BS in accounting at Eastern State this May, and have found cost accounting and the related issues in operations management an area in which I am confident I would enjoy working.

    I realize that I do not offer the precise experience stated in your advertisement. However, my experience as a teacher and high school coordinator has helped me develop many skills, including organizing my work, directing others in their work, solving problems, and delegating duties to others. This experience should help me quickly adapt to a job as a cost accountant at Four Diamond.

    While at Eastern State, I have taken several courses devoted to accounting management technology. I have also used current software programs to complete much of the classwork in accounting, finance, business policy, statistics, and operations management. I feel comfortable with the Microsoft Office suite, especially Excel, Word, and PowerPoint. I feel I can also adapt to whatever programs you use.

As I hope you will agree, I fit your requirements closely. I look forward to discussing this opportunity with you in greater depth. I am available via email at jeremiah.jones@esu.edu and via phone at 904-386-4163 after 4:00 p.m. on weekdays.

FOR COLLABORATION

At your instructor's request, form teams of three students to prepare a brief report on job opportunities in user experience design (UXD). Use different job posting sites and professional organization websites, including the U.S. Bureau of Labor Statistics (www.bls.gov), to research what UXD is, its career outlook and projected growth/decline, and other relevant information. Find out what skills and qualifications are required for an entry-level UXD position.

REFERENCES

Smith Diaz, C. (2013). Updating best practices: Applying on-screen reading strategies for résumé writing. *Business and Professional Communication Quarterly, 76*(4), 427–445.

Watson, M. (2019). Using professional online portfolios to enhance student transition into the poststudent world. *Business and Professional Communication Quarterly, 82*(2), 153–168.

13 Proposals and Grants

For more about
problem definition,
see Chapter 5

In a proposal, you offer to provide a product or service or to do some kind of work to solve a defined problem. It can be a simple one-page document written by one person for one reader. It can be a multivolume document composed by a team of 300 people and read by 20 or 30 evaluators. It can be something in between. Broadly conceived, proposals document ways to make the world a better place. They are major tools in business, in research, and in the classroom.

You may also write a proposal when you apply for a grant—that is, funding support for your individual research or a collaborative project. Government agencies and non-profit organizations are the major sources of grants. In this chapter, you'll learn strategies based on design thinking that will help you prepare and deliver proposals and grant applications.

PURPOSE

Your purpose in writing a proposal is to show how, in a particular situation, you can solve a problem for someone who needs that solution or carry out a sponsored project. The situations can be complex. In this section, you will learn about three common situations that shape the writing of a proposal: bid, implementation, and research.

Bid

In the simplest situation, you provide a routine response to a well-identified problem. The product or service you're offering comes off the shelf, and the major differentiating item is usually the price. Your proposal, often called a bid, may contain a few lines of text and columns of numbers indicating costs. Once the bid is accepted, the document becomes a binding contract; for example:

- Distributors of cleaning supplies write bids to provide their products to corporations, restaurants, and other organizations.
- Furniture companies write bids to sell equipment to universities.

Less routine bids tailor a known procedure, product, or service, such as business process software or sophisticated navigational systems, to a particular customer.

DOI: 10.4324/9781003093763-17

In these circumstances, the bid may contain extensive discussion about how the item will be made to work for the customer:

- Defense contractors write proposals to governments worldwide to build or recondition ships and planes.
- Geotechnical engineers write proposals that describe the method they would use to determine soil conditions on a building site.
- Consultants write proposals that show how they would prepare an environmental impact statement for a project.

Implementation Proposal

Beyond a simple bid, you may find yourself proposing to design and implement a new system or process. No off-the-shelf approach will do. Your document must account for the problem, as you see it, and the way to improve the situation. The box on "Implementation Proposal" reproduces an implementation proposal. Here are some other contexts for such proposals:

- A nurse proposes the implementation of an improved triage system in an emergency room.
- An engineer proposes an oil filtering system to recycle old oil and, thus, save money in both purchasing and disposal.
- A technician proposes a new system for recording the food intake and weights of laboratory animals.
- A technical communicator proposes a new system for transitioning to a structured authoring system.

Research Proposal

You may have written a brief proposal in an introductory academic writing course to describe a research problem and how you would explore or address it. Such a research proposal falls into the third proposal situation.

In a research proposal, you define a problem, one that may not have been well recognized or defined. Then you describe a method for finding a solution and convince the reader that the method is likely to succeed. The audience may be a supervisor, a client, a sponsor, or a professor. Figure 13.2 reproduces a student research proposal.

Major government funding agencies, such as the National Science Foundation and the National Institutes of Health, and major private not-for-profit organizations, such as the Robert Wood Johnson Foundation and the Rockefeller Foundation, award contracts and grants for research in response to proposals that compete for such money. The funding agency usually establishes a theme or priority for the kind of research it seeks: identifying the causes and treatment of diseases, developing ways to prevent crime, or coming up with innovative techniques for handling waste are some representative priorities.

Date: 30 October 2019
To: Chuck Blackman, Manager
From: Peter Sanchez
Re: New System for Processing Inventory Shipments

This is a proposal to implement a new system for processing inventory shipments. The proposal derives from my own observations and from conversations with several other employees.

Problems with the Current System
Currently, shipments are not being opened and stored promptly. This delay in the availability of inventory has meant that employees find it difficult to meet customer requests. Inventory that remains in boxes or is not stored properly cannot be located or sold. In addition, employees also do not know the current status of merchandise. Third, stockpiles build up—when one shipment is not stored as the next one is received—it is twice as hard to catch up on the work. Inventory fills the aisles of the stockroom and makes it hard to move around and do one's job. These problems are leading to an even more serious one: a decline in employee morale and customer service quality.

Objective of the Proposed System
The main objective of the proposed system is the timely processing of merchandise shipments. Such processing depends on:
- Separating the inventory process into discrete tasks: receiving, opening, verifying, and storing.
- Delegating authority to one individual who is responsible for the process.
- Designating certain employees as specialists in the process.

Implementation of the Proposed System
The proposed system reflects comments from current employees as well as my own observations of the process as it is now performed.

Except in rare situations, shipments arrive on Monday and Friday. For those two nights:
1. Schedule a larger workforce. Three additional employees should work as processing specialists on evenings when shipments arrive.
2. Assign the same individuals to work those evenings so that the process becomes routine for them.
3. Separating shipment processing into discrete tasks. These tasks are receiving, opening, verifying, and storing. All three people can help to bring the shipment into the stockroom. They then separate the boxes on the basis of their content: shoes, apparel, and hard goods. Each person takes one of the three types of items and verifies the shipment against the order. When the shipment is verified, each person can properly store the merchandise.
4. Delegate authority to one of the three specialists. In that way, one person can resolve any disputes without due questioning.
5. Designate processing as a responsibility. Provide a specific timetable for each the processing specialists.

Benefits of the New System
The cost of the extra employee time on shipment evenings is more than outweighed by the benefits of the new system. These include:
- A more organized stockroom
- More easily accessible inventory
- Better customer service
- Enhanced sales
- Enhanced employee morale

For all these reasons, I'd recommend that we implement the proposed system as soon as possible. I look forward to your feedback and decision.

13.1

Implementation Proposal

Deathcare on the Web

An Exploratory Analysis of Funeral Home Web Design Conventions

Background: The Proliferation of Funeral Home Websites

North American deathcare began as a familial tradition performed inside the home and has since transformed into a multi-billion-dollar industry (Farrell, 1980). At the core of the modern deathcare apparatus is the funeral home, a community-based commercial establishment responsible for carrying out many of the day-to-day activities of deathcare work. As a response to the deathcare industry's rapid professionalization and commercialization, funeral homes have gradually evolved to become more technologically and digitally mediated, thus leading to the proliferation of funeral home websites.

Funeral home websites typically fulfill several key professional functions. First, they are digital representations of the firm's brand, which is often built on epideictic values of respect, trust, and dignity. Secondly, the funeral home website is an informative hub for all-things-deathcare, providing users with various details relating to the planning and execution of funeral services. Further, these websites may help facilitate public mourning by sharing obituaries and other memorial tributes, as well as hosting public forums to encourage reflection and community engagement. In many ways, then, the funeral home website is a rich professional and cultural site that not only reflects values of the profession, but also the inherent tensions between tradition and innovation.

Although there is a rich body of existing research focusing on web design and usability (Cyr et al., 2009; Faisal et al., 2017; Pengnate & Sarathy, 2017; Song & Zahedi, 2005), this scholarly interest has not yet been extended to the domain of deathcare. Therefore, we lack an understanding of basic funeral home web design conventions and how those conventions may adhere to or perhaps even disrupt design conventions typically associated with business, professional, or other e-commerce websites.

Purpose: Unearthing Funeral Home Web Design Conventions

My exploratory pilot project seeks to identify some of the basic design conventions of funeral home websites in order to better understand how these conventions may reflect broader characteristics of the deathcare industry at-large. The following questions guide the study:
- What are the basic design conventions of funeral home websites?
- How might these conventions function rhetorically?

Methods: Interpretive Content Analysis

Due to the exploratory nature of this project and the time constraints imposed by the class, I will limit my research to Mississippi funeral home websites. Using a database of licensed funeral establishments in the state, I will visit each funeral home website available and conduct an interpretive content analysis (Neuendorf, 2017) focused primarily on the following main design variables: basic layout and organization, typography, color, use of

13.2
Student Research Proposal
Source: Courtesy of Wilson Knight

images, and overall branding. Because of the formative scope of the project, I will also collect data on any significant emerging variables relevant to the purpose of this study. To streamline my data analysis and ensure reliability, a standard coding scheme will be used to analyze each variable.

After using a primarily quantitative approach to content analyze each Mississippi funeral home website, I will look for patterns emerging from the data and begin to consider how these patterns might reflect broader values of the deathcare profession. Likewise, I will look for any differences that emerge in design conventions based on external factors such as corporate/private ownership, rural/urban location, and caseload. Using a mostly-quantitative approach to content analysis will allow me to analyze trends more meaningfully across the dataset, while also developing a useful framework that can be replicated in future studies of professional website design.

Anticipated Findings & Implications: Funeral Websites as Sites of Tension

First, I expect many funeral home websites will reflect similar conventions, such as use of neutral color palette, formal typography, and basic organizational structure. These trending conventions will most likely align with many of the fundamental values of the deathcare profession, so identifying the common conventions of Mississippi funeral home websites may be a productive first step to identifying a more national trend amongst deathcare professional websites.

While several design conventions may be similarly applied across funeral home websites, I do expect to find some tensions between modern design conventions and the profession's commitment to tradition and decorum. These tensions may help prompt us to develop future design recommendations that more meaningfully apply basic design principles without compromising the integrity or character of the deathcare profession.

Findings from this study may be used to generate future research on design conventions within deathcare and other unique professional domains. Likewise, results may be useful for deathcare professionals or designers who are interested in developing a more intentional awareness of their web presence through the lens of design. Thus, this research will contribute to scholarly conversations taking place in the fields of communication studies and death studies, as well as to practitioners working in aligned fields.

References

Cyr, D., Head, M., Larios, H., & Pan, B. (2009). Exploring human images in website design: A multi-method approach. *MIS Quarterly, 33*(3), 539-566.

Faisal, C.M., Gonzalez-Rodriguez, M., Fernandez-Lanvin, D., Andres-Suarez, J. (2017). Web design attributes in building user trust, satisfaction, and loyalty for high uncertainty avoidance culture. *IEEE Transactions on Human-Machine Systems, 47*(6), 847-859.

Farrell, J. (1980). *Inventing the American way of death, 1830-1920.* Temple University Press.

Neuendorf, K. (2017). *The content analysis guidebook* (2nd ed.). SAGE.

Pengnate, S., & Sarathy, R. (2017). An experimental investigation of the influence of website emotional design features on trust in unfamiliar online vendors. *Computers in Human Behavior, 67*, 49-60.

Song, J., & Zahedi, F. (2005). A theoretical approach to web design in e-commerce: A belief reinforcement model. *Management Science, 51*(8), 1219-1235.

FRAMING PROBLEMS USING A POINT-OF-VIEW STATEMENT

Broadly conceived, proposals articulate ways to understand a problem and identify solutions to solve it. Before you go into the problem-solving mode, however, one very crucial step is to *frame* the problem based on the situation that contextualizes it (see more in Chapter 5). In design-thinking exercises, a point-of-view (POV) statement is used to perform this kind of framing by focusing on stakeholders of the problem.

Create a POV statement by combining your knowledge about the audience or consumer you're addressing, including their needs and insights, which you can gather through initial research. A POV statement follows this structure:

[Audience] *needs to* [Requirements] *because* [Insights]. *However*, [Gap/Problem].

For example:

Adult persons who live in the city *need* access to shared transportation 1–4 times for 10–60 minutes per week *because* they believe it is cheaper, and more environmentally friendly. *However*, these transportations are not readily available in Chicago.

SOLICITED AND UNSOLICITED PROPOSALS

Within each of the above contexts, a proposal may be solicited or unsolicited, depending on who recognized the need or problem first. If it's the client, customer, sponsor, or professor who sees the problem and requests a solution, the result is a *solicited proposal*. When writers think of the problem on their own, they create an *unsolicited proposal*. In practice, however, as you'll see, this neat distinction breaks down. When you identify a problem that needs to be solved, you are likely to be more successful in proposing a solution if you can talk with the potential audience, so that they then solicit the proposal.

Solicited Proposal

A solicited proposal responds to a direct request. That request may come from your professor in an academic setting. Outside the classroom, government agencies, businesses, and individual clients solicit responses to solve a need or problem. That solicitation may circulate in conversations or, more formally, in a classified advertisement or document called a request for proposal (RFP), request for bid (RFB), or request for qualifications (RFQ). RFPs may be as short as a paragraph or as long as a volume. The www.grants.gov site provides links to U.S. grant-making agencies along with tips for developing proposals. The following excerpt is a National Science Foundation (NSF) solicitation that relates to integrative computing systems (note the acronyms):

Cyber-physical systems (CPS) are engineered systems that are built from, and depend upon, the seamless integration of computation and physical components. Advances in CPS will enable capability, adaptability, scalability, resiliency, safety, security, and usability that will expand the horizons of these critical systems. CPS technologies are transforming the way people interact with engineered systems, just as the Internet has transformed the way people interact with information. New, smart CPS drive innovation and competition in a range of application domains including agriculture, aeronautics, building design, civil infrastructure, energy, environmental quality, healthcare and personalized medicine, manufacturing, and transportation. CPS are becoming data-rich enabling new and higher degrees of automation and autonomy. Traditional ideas in CPS research are being challenged by new concepts emerging from artificial intelligence and machine learning. The integration of artificial intelligence with CPS especially for real-time operation creates new research opportunities with major societal implications.

The CPS program aims to develop the core research needed to engineer these complex CPS, some of which may also require dependable, high-confidence, or provable behaviors. Core research areas of the program include control, data analytics, and machine learning including real-time learning for control, autonomy, design, Internet of Things (IoT), mixed initiatives including human-in- or human-on-the-loop, networking, privacy, real-time systems, safety, security, and verification. By abstracting from the particulars of specific systems and application domains, the CPS program seeks to reveal cross-cutting, fundamental scientific and engineering principles that underpin the integration of cyber and physical elements across all application domains.

Proposals for three classes of research and education projects—differing in scope and goals—are supported through the CPS program:

- **Small** projects may request a total budget of up to $500,000 for a period of up to 3 years. They are well suited to emerging new and innovative ideas that may have high impact on the field of CPS. There is no deadline for Small projects.
- **Medium** projects may request a total budget ranging from $500,001 to $1,200,000 for a period of up to 3 years. They are well suited to multidisciplinary projects that accomplish clear goals requiring integrated perspectives spanning the disciplines. There is no deadline for Medium Projects.
- **Frontier** projects must address clearly identified critical CPS challenges that cannot be achieved by a set of smaller projects. Furthermore, Frontier projects should also look to push the boundaries of CPS well beyond today's systems and capabilities. Funding may be requested for a total of $1,200,001 to $7,000,000 for a period of 4 to 5 years. Note that the Frontier projects have a specific deadline.

Source: National Science Foundation, Computer and Information Science and Engineering (CISE). Retrieved from https://bit.ly/3tDjeya

Unsolicited Proposal

An unsolicited proposal results from the writer's own perception of a problem. You have to convince the reader there is a problem before you can begin to write about solving it. This situation is particularly risky within organizations because you may point out a problem to the very person who created it. You also risk your own time in pursuit of a problem no one may see as such. But such a proposal may be just what you need, like the implementation proposal example shown in the box on "Implementation Proposal."

THE STRUCTURE OF A PROPOSAL

Proposals propose a direction and provide a structure for doing so. The purpose of a proposal is to show how, in a particular context, a problem could be addressed.

Introductory Elements

Proposals generally share a common document structure, which can be modified to suit specific circumstances. The following are the typical elements in a proposal.

For more about executive summaries or abstracts, see Chapter 15

- *Executive summary*: Provides an overview of the proposal with highlights of the problem addressed, proposed activities, and project deliverables.
- *Introduction*: Summarizes your proposed activity by stressing benefits to the reader. Provides the context by briefly restating the problem (if the proposal is solicited) or establishing the problem's background and urgency (if unsolicited). Overviews the content and plan of the proposal for a wide readership.
- *Statement of problem*: Identifies the problem necessitating the proposed work, with enough detail to make the problem clear to the reader. Provides a review of relevant literature if that is needed to establish the significance or dimensions of the problem.
- *Objective(s)*: Lists the specific, measurable outcomes you plan to accomplish. Explains the solution you will propose. In a research proposal, shows that your work will contribute to the theme established by the funding agency or the assignment made in class.

Methods/Methodology

This explains either how you will implement your solution (if the solution is known) or how you will conduct research to support your hypothesis about the solution (if it is unknown). It ties the activities directly to your objectives and convinces the reader that your approach is reasonable, suited to your resources in people and facilities, likely to succeed, and better than the competitor's. It may include a review of literature to show how the proposed methods derive from but improve on those of other approaches and are viable. It notes compliance with any federal, state, or local laws in undertaking the work.

Project Management

In a collaborative proposal effort, this shows how the team will be coordinated or managed. It gives profiles of key staff along with the extent of their participation in the proposed project (extended biographies may be included in an appendix). If your project deals with primary data (collected through firsthand sources such as surveys, interviews, etc.), it explains how data collected will be stored and shared among collaborators. It also describes ethical considerations and compliant practices in sensitive data management.

Schedule

This places your implementation or research activities on a timeline and convinces the reader that the timeline is realistic. It serves as the proposal-at-a-glance.

Justification

The justification answers the question, "Why you?" It provides your track record of relevant accomplishments and qualifications and assures the reader that adequate staff and facilities are available to carry out the project as outlined. It describes labs or field sites, specialized equipment, and other digital resources matched to the tasks and convinces the reader that you are prepared to carry out the project.

Budget

This assigns monetary values to all activities and resources mentioned in the proposal. It is often read first as, in effect, the quantitative abstract of the proposal. It divides the total budget into categories with proper budget justifications.

For more about persuading versus explaining, see Chapter 8

STRATEGIES FOR PERSUASION

As proposals aim to define problems and identify solutions, they should be composed to persuade readers to adopt the proposed idea.

Introducing the Context

Because many proposal evaluators decide after reading only the first paragraph or two whether to read further, pay particular attention to your introduction. The introduction sets the stage. Spend more time on the problem under review when the proposal begins at your own initiative than when you are asked to solve a problem the reader already knows about, perhaps all too well.

The following paragraphs open a proposal from a team of researchers studying climate change. Note how the authors practice design thinking, calling upon reader empathy:

> Minnesota has a mercury problem. Eight percent of infants born in the Lake Superior Basin in Minnesota have mercury concentrations in fetal cord blood that exceed

human health standards, apparently related to fish consumption by their mothers. In addition, 95% of the stream reaches and lakes assessed for mercury in fish have been listed as impaired, posing a threat to human and environmental health. This problem will be exacerbated if increased temperatures expected with climate change increase decomposition in peatlands, which in turn will release at least a fraction of the vast stores of mercury and sulfur they have accumulated from both anthropogenic and natural sources in atmospheric deposition. This release would directly increase mercury and sulfur concentrations in streams and lakes and also enhance the formation of methylmercury, the most toxic form of environmental mercury [. . .]

We propose to measure the effects of increased temperature on the release and fate of mercury and sulfur from peatlands and to use that knowledge to predict their specific impacts on Minnesota's aquatic ecosystems. Both of these elements can be released to surface waters, with additional impacts on local aquatic ecosystems that are already of concern, or volatilized to the atmosphere to be more widely dispersed. Consequently, we will analyze and predict both pathways for each element.

If you write the proposal as a memo, the heading provides a preview for the reader, who then comes to the memo itself with its topic in mind ("Implementation Proposal" and "Student Research Proposal" boxes). The implementation proposal shows in the introduction how the writer turned his own observation of a problem into a situation in which his supervisor requested the proposal. The opening is short, appropriate for a business document. The research proposal addresses the students' professor, who set several conditions for the final project to be detailed in the proposal:

- Identify and address an audience (not the professor)
- Choose a problem related to the environment
- Use both empirical and document-based research methods.

The introduction immediately responds to the first three items and sets up later detailed discussion of all four conditions.

Defining the Problem and Urgency

For more about the types of problem, see Chapter 5

The process of preparing a proposal starts with the identification and analysis of a problem, a situation ripe for change. If you have been assigned a problem, or if you are responding to an RFP, comb the assignment or the RFP for every aspect of the problem to be addressed. On that basis, decide if you indeed have what it takes to write the proposal. You can't turn down a classroom assignment, but you have that option in other contexts. In the classroom, too, select problems that are workable. Do you have the skills to study the situation or the solution to match the problem? Can you comply with the specifications? Do you have time to prepare the proposal?

As you gather information, make sure you are dealing with only one major problem per proposal. To sort through the problem, read articles in pertinent journals or magazines or search documents available online that are applicable to the topic. Such reading will aid you in understanding the dimensions of the problem and establish its importance. The inventory proposal delineates four causes for the delay in processing shipments at the sporting goods store.

The following problem statement anchored a student's proposal for research titled "Examining the Perceptions and Impact toward International Students from Teaching Assistants in Undergraduate, Non-Major Biology Laboratories":

> Underrepresented and minority groups within higher education face a lack of equity across STEM disciplines. Recent literature has displayed gender-based, racial, and ethnic disparities, among others, to be prevalent issues causing lower performance, participation, and rates of retention. While issues such as the gender gap are currently being addressed, there are many other identity-related barriers students are facing in the classroom. One such issue that has been overlooked is that of international students facing isolation due to language barriers, the lack of relationships with host nationals, racial discrimination, and poor integration. Social isolation, and developing a social network, is the primary issue of adjustment for international students; however, this issue begins in the classroom when students are not given the support they require. Coping behaviors exhibited by these students within the classroom are comprised of little class participation, a lack of asking clarification questions, and sitting and studying with other international students (thus further delaying integration).

Describing Your Objectives and Activities

How do you propose to deal with a problem? Every proposal must have a core idea, its major sales pitch. This becomes most evident in the objectives segment. The objectives are outcomes that can be measured to tell you the problem is solved; to ease that measurement, objectives are usually stated in list form. They set the scope of the project. In an implementation proposal, it's the selling feature—such as the "timely processing" of inventory. In a proposal for research, the core idea is your research questions or your hypothesis about what you'll find when your research is completed.

How will you meet these objectives? In an implementation proposal, the methods segment shows the steps you'll take, as in the segment headed "Implementation of the Proposed System" in the "Implementation Proposal" box. In a research proposal, you describe your method for testing the hypothesis or addressing a research question. Your hypothesis may be your proposal's major selling point, but you may also find that your method is unique, and thus it holds the major appeal. Set out the tasks you need to perform, perhaps divided into phases.

The more detailed and specific your description, the more convincing you'll be. Readers evaluating your proposal will measure your activities carefully to determine if the scale or scope is appropriate and if the plan is realistic and likely to succeed.

You also need to show readers that you have neither underestimated nor overestimated the difficulty of the work or the time it will take to complete it. To that end, you provide a schedule of tasks and their completion dates, usually in visual form (see Figure 13.3). In a formal proposal for a collaborative project, you also provide information about team member credentials. Depending on the length and

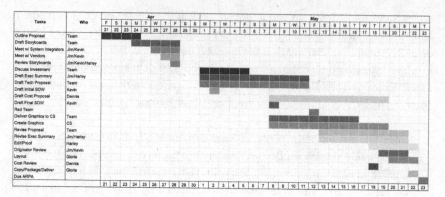

13.3
Visual Schedule

complexity of this information, you may include it as part of the methods or justification section or in a separate segment of its own.

Showcasing Your Qualifications

Why, in the end, should the reader accept your proposal? Your major selling point may be your core idea, your method—or you. Particularly in high-context cultural settings, your reader's trust in your professional reliability may be enough to convince. As your personal résumé supports what you can do on the basis of what you have done, so a proposal persuades on the basis of the proposer's track record and character. Avoid simply boasting—"We make the finest widgets in the world." Instead, be concrete: "Our widgets have operated for more than 25,000 hours without a single failure." Give the reader clear and compelling reasons to choose you. Detail your track record on similar problems. Also provide the organization's credentials and the available support system—in people and facilities. These resources must be adequate to complete the tasks successfully. If you lack appropriate expertise or resources, you may arrange a collaborative effort with another organization that provides those resources.

For more about cultural settings, refer to Chapter 1

Projecting Costs and Return on Investment

How much will your project cost? In the budget, you identify the resources needed to carry out what the document proposes. Professors may not expect budgets in student proposals written as classroom exercises, but evaluators in the workplace often start their reading with the budget. They may see the budget as the distillation of the entire proposed project and a test of that project's realism and worth. As elsewhere in the proposal, closely follow any direct guidelines in preparing the budget. Calculate salaries or other payment to personnel as requested: by hour, by day, by month, or by some other unit.

MANAGING A COLLABORATIVE PROPOSAL EFFORT

Collaborations are common in proposal writing, especially with a complex or large-scale problem or solution (or both). You may find it helpful to join forces with people who share your expertise, or represent other kinds of expertise, to develop an effective proposal. A typical proposal from a defense contractor may require a large team, whereas start-up projects may require only a small team of two to three collaborators. At any rate, proposal teams need to achieve consensus on core concepts, share tasks, manage communication, and consider cross-cultural collaborations.

Agreeing on the Core Concept

Whether your proposal team is large or small, your first responsibility is to agree on the proposal's core concept. As in any collaborative effort, proposal teams can run into trouble when members approach the project with varying ideas about the proper outcome, varying levels of enthusiasm, and varying commitment to a task that is often not their main responsibility. A clear sense of purpose and good leadership help the team overcome these problems. They also prevent team members from duplicating each other's work or working at cross-purposes, one member, for example, preparing extensive demographic data on a region that other members decide isn't relevant to a proposed survey. Sorting through core ideas and tasks early in proposal preparation and debugging potential solutions mean fewer rewrites and less overtime fixing a poor proposal.

Assigning Tasks

As they carry out the tasks required to fulfill the proposal's purpose, team members usually form into three subgroups, each group performing a major proposal function: marketing, technology, or communication. Often, the groups perform their functions in a linear sequence. The marketing group, for example, begins the process by initiating or maintaining contact with the client or customer and then identifying the problem to be addressed. Marketing next turns the problem over to technical experts who determine whether the company is the right one to solve it and, if so, how to solve it. Finally, the communication specialists write the proposal. Several proposal experts, however, say the process works more effectively when representatives of all three functions collaborate throughout the processes. They, thus, can raise and resolve conflicts in advance of the competition external to the company.

In this coordinated approach, everyone contributes to the discussion and everyone writes. The experts at each function, however, take primary responsibility for developing certain kinds of information. The marketing experts, for example, analyze the context. They determine the customer's values and needs as well as the *critical success factors*—that is, how the customer will measure the success of the project. To do this, they monitor potential opportunities, analyze the competition, and develop a strategy for tracking the customer. The technical experts provide the major content of the document. Depending on the proposal, that group may include scientists, engineers, lawyers, accountants, or other professionals. Some establish the detailed statement of the problem—the core idea, product, or service that will solve the problem—and the method may require elaboration in a step-by-step description of a procedure. Others, experts at financial matters, create the budget. If the proposal requires extensive legal commentary, then lawyers may prepare that information. Communication experts interweave the marketing and technical efforts to articulate the clear, persuasive message each proposal must convey. They review the entire proposal to make sure it is well organized, easy to read, and well suited to the customer. They also supervise the design and production of the proposal.

For more about proposing for business situations, see Chapter 14

Managing Team Communications

Team members meet regularly to make sure the work is proceeding appropriately. They may also keep in touch through frequent email or phone contact. As a group, they monitor their work. In addition, the leader keeps close tabs on progress. Two specific tools for monitoring are a schedule and a critical review.

Schedule

Proposals are often written under near-panic conditions. To reduce the sense of panic and foster good thinking, even when deadlines are tight, the team needs to plan its activities. That plan is made concrete in a schedule that establishes each task and names the person responsible for that task. Begin with final deadlines and work backward, dividing the time available by the tasks to be performed. Note, by the way, that most final deadlines are firm: if your proposal isn't received by that date, it isn't considered. Figure 13.3 shows the schedule for preparing a complex proposal in the defense industry.

The schedule helps coordinate the team effort by reminding everyone of what has been done and needs to be done and providing the leader with information about any need to redirect the work or add to or reduce the team. Several schedule forms are common, and many of them are available in computer programs that automatically plot and update the sequence.

Critical Review

In some intensely competitive situations, the proposal team may also monitor the quality of its work through a critical review. For the review, similar to those that architecture students, for example, participate in at the end of a term, a panel of experts questions the team and evaluates the proposal. A common form in the defense industry is called a "Red Team Review." The Red Team consists of company experts who take the role of the customer or client and look for any weakness in content or presentation. Red Team comments then help the team revise its draft.

Collaborating on Cross-Cultural Teams

Collaborating on a multinational proposal team requires extensive upfront planning as well as patience. The U.S. style may often be competitive, aggressive, detailed, and urgent. But many Europeans chafe at the conventional persuasive style of U.S. proposals. They may not see as necessary the close reading of an RFP. They are less inclined to negotiate detailed statements of work. And they may take deadlines less seriously than Americans. With short-term thinking and an emphasis on getting things down immediately, American team members may resent delays and simply write over the contributions of their European teammates. Such behavior erodes team trust.

In addition to technical concerns, the cross-cultural team must also address issues such as currency exchange rate differences, export and import licenses,

For more about
the purposes of a
business plan, see
Chapter 14

percentage of work to be conducted in each country, and patents and intellectual property rights. If a joint venture is to be established, it may require an international business plan, legal incorporation, and regional headquarters. Translating the executive summary of a proposal, or an entire business plan, can add other difficulties—because many terms that are standard to Americans can be misleading or offensive in another context, because equivalent terms are not always available or may be debatable in another language, because the fast pace of writing may not be matched by similar speed in translating, or because any British members of the team may object if American English prevails.

To offset these difficulties, the American members of a multinational proposal or business planning team need to spend more time talking with their teammates and set a pace that allows for that process. All members of the team, or at least representatives of each national partner, should sit in on discussions of draft components of the document. Such open discussions reveal variations in the way different companies and their cultures do the same thing, along with sensitivities toward national issues, and helps to resolve differences in emphasis, terminology, and values. It also helps to develop a shared glossary and style guide so that the resulting proposal or business plan displays consistent usage and speaks with one international voice.

DELIVERING THE PROPOSAL

How your proposal reads and looks conveys an important message about how your proposal will work. Review your draft for style and design.

Style and Expectations

Assess carefully your audience's values and way of seeing things and gear your expression accordingly. Nigel Greenwood, a technical writer in France, notes, for example, that French proposal writers consider it only respectful to assume that readers can work out the details on their own. They, thus, emphasize imagination and elegance of language over explicit, factual content and cultivate ambiguity rather than clarity. In addition, the French tend to prefer abstract ideas and to rely on authority, whereas people in Anglo-Germanic cultures like that in the United States focus on more prosaic issues such as who will do the work and how the solution will be implemented. So, French proposal writers may use long, sophisticated sentences that digress with abandon and follow several thoughts at the same time. An evaluator in the United States would expect a more rigorously linear structure.

In addition, be careful to use the audience's specialized language and, at the same time, avoid your own code words that the audience might not understand. Check the RFP, for example, for key terminology and make sure you use those terms in your response. Avoid synonyms (calling something a "transmitter" in one place and a "beacon" in another) because they may confuse readers. Spell key words the way the audience spells them, even if that's different from your organization's style.

To match the needs of most U.S. evaluators, be positive and bold with verbs, avoiding passives ("what is to be done") that cause ambiguities. Instead, note that specific people—*I* or *we* or a particular person whose name appears as the subject—will take specific actions. Differentiate among such hard-worked words as analyze, determine, fabricate, study, define, and characterize. Use the simple present tense for general descriptions and the future tense for specific statements of future activity:

- This proposal outlines a strategy for analyzing three parcels of land in Sussex County.
- We will conduct an on-site investigation.

Design and Presentation

No one would approve a proposal or business plan on looks alone, but a professional-looking document predisposes the reader toward reading and toward a good opinion of you. So, pay attention to design as you draft the proposal and especially as you develop the package you'll send to your client, sponsor, supervisor, or professor. As with a personal résumé, good design enhances both readability and credibility. In Figure 13.4, you see segments of a well-designed proposal to the U.S. Department of Defense. Its simplicity reinforces a message of cost-effectiveness to readers who might be suspicious of glossy treatment. Use lists and headings within the text to aid people who skim proposals and to reiterate your core sales message.

For more about résumés, see Chapter 12

Digital Delivery

Many research and implementation proposals are delivered digitally today. You are likely to create and submit your materials within an online content management system where you're allowed to revise your materials until the proposal deadline.

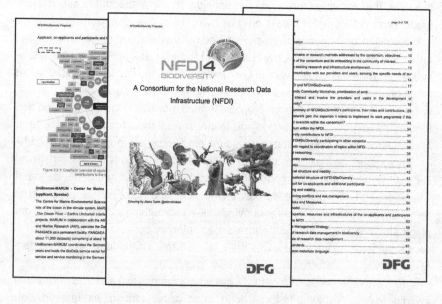

13.4
Preliminary
Segments of a
Proposal

Source:
NFDI4BioDiversity—
A Consortium for the
National Research Data
Infrastructure (NFDI).
Retrieved from
https://bit.ly/2QIriPk

As project director or proposal writer, you should familiarize yourself with the submission portal by reading or viewing the submission guidelines, creating relevant application login credentials, and setting up project routing and tracking accounts within your institution. Although these may seem trivial as the intermediary steps in the proposal process, they can most certainly be determining factors for the success or failure of a proposal project.

PASSING THE TEST

A proposal has one central message: I (or we) can solve your problem or improve your situation. To convey that message:

- Identify the problem or situation
- Explain your solution or improvement or detail your method for finding a solution
- Justify why the client or customer should choose you.

Conveying that information convincingly is the role of the proposal. Your proposal will meet a critical audience, and many proposals fail that review. One evaluator for a government agency said she rejected 85 percent of the proposals sent to her on the basis of unconvincing problem statements. Other reasons for rejecting a proposal include the following:

- The solution doesn't match the problem.
- The research idea is trivial or overcooked.
- The methodology is inappropriate, not well thought through.
- The amount of work proposed is unrealistic for the time frame.
- The proposal doesn't comply with the RFP's stated specifications for content, format, or schedule.
- The proposed budget is too high.
- The proposal includes special provisions that conflict with the stated policy of the client or customer.

For more about professional correspondences, see Chapter 18

When a proposal is rejected, it's a good idea to find out why. Government agencies, for example, usually arrange such post-review, or debriefing, conferences. The conference can help you shape future proposals.

When the proposal is accepted or approved, you undertake the work. The proposal turns into a binding contract specifying the amount and conditions of the work. As part of that contract, you may write progress reports at predetermined intervals. If the work requires research or development, its end may be marked by a final report and, perhaps, an article for colleagues.

For more about formal reports, see Chapter 16

A well-written proposal provides a good framework for writing the final report. But the two are not the same. Although it includes information about methods, the final report emphasizes results. Where the proposal looks forward, the final report looks back, and, thus, you approach the two documents differently.

ALWAYS BE ITERATING

While you may be bound by the contractual terms and scope of work in your approved proposal, you should remain open to potentially better solutions or methods in executing your proposed activities. An iterative approach to problem solving means creating gradual improvements in your work. Start by carrying out what you have planned in the approved proposal and, as you progress, identify areas for incremental improvements and make those changes along the way.

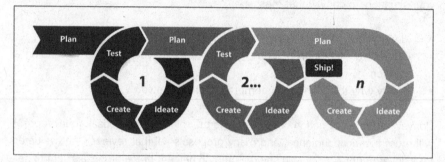

13.5
Iterative Approach
to Problem Solving

Whether your proposal is a one-page memo or a multivolume collaborative venture, remember it is a sales document. As such, it should be accurate and promise only what you can deliver. It should meet the stakeholder's needs. And it should showcase your good ideas. Bear in mind, however, that good ideas are often contextual and dependent on existing resources. Better resources often emerge over time. Don't ignore them. Incorporate them in your thinking. Keeping an iterative process and mindset, as encouraged by design thinking, can help you identify incremental improvement opportunities. Figure 13.4 shows this iterative approach to creating gradual solutions.

Proposals and grant applications are tools for solving defined problems. In this chapter, you have learned three common situations for proposal and grant writing—bid, implementation, and research. Use the recommendations provided for reading and responding to RFPs in your own field of work. Apply the chapter's guidelines in designing proposals that address the specific requests as well as managing collaborative efforts. When presenting to clients, design a package that is visually pleasing and persuasive.

CHECKLIST: PROPOSALS AND GRANTS

1. Write a proposal to offer a solution or method to solve a problem
   Write a bid to offer a routine response to a known problem
   Write an implementation proposal to design and implement a new process
   Write a research proposal to investigate a difficult problem with a viable method

2. Read solicitations/RFPs or look around for problems not previously identified
3. Design the proposal to match the request or your analysis of the problem
   - Statement of problem and benefits
   - Objective(s)
   - Methods and activities
   - Management of project
   - Schedule
   - Justification
   - Budget

4. Manage a collaborative effort by
   - Generating and agreeing on the core idea/selling point of your proposal
   - Assigning specific tasks and scheduling them appropriately
   - Monitoring progress to ensure quality, responsiveness to the sponsor/funder, and timeliness

5. Deliver the proposal in an attractive package via the right medium
6. Follow through, both when your proposal is accepted and when it is not

---

EXERCISES

1. Justify your request for funds to carry out a research project that might make you eligible for a scholarship or an honors program. First, obtain the requirements for the project, then write a response that conforms to those requirements.
2. In a memo to the chair of your department, propose a revised set of requirements for graduation from your major. Be sure to indicate what the present requirements are. Then, describe the criteria you think the requirements should meet and note why the present requirements fall short of—or exceed—that standard. Finally, describe your proposed changes.
3. Write an email to your college library proposing that the library should subscribe to three (or more) academic journals that are not currently in its collections. Explain the value of these journals to students and give the publication data and subscription prices.

FOR COLLABORATION

At the beginning of a team reporting project, submit a proposal to your instructor that outlines the problem you will address, the audience for the report, your procedure (including the tasks you'll undertake and how you'll manage team communications), your schedule, and the intended outcome. The proposal should help you coordinate teamwork and allow the instructor to evaluate your approach and suggest any corrections to it if necessary.

14 Business Plans

To foster innovation and continuous improvement, business organizations produce business plans, a genre of technical communication that benefits in particular from design thinking. This chapter teaches you how to compose different business plans to address different organizational needs.

PURPOSE

A business plan is a document that describes the nature of a business, its products and services, sales and marketing strategy, financial background, and operations detail. While needed for new companies, every company should have a business plan to guide specific endeavors. Strong business plans provide clarity and specificity in terms of mission, vision, service/product area, and targeted audience, as well as viable solutions to existing problems in a company with a feasible budget and timeline. In the following section, you will learn that different situations require specific business plans.

SITUATIONS

In order to design a persuasive business plan, you need to understand the demands of the business situation or problem. Business plans are written to guide organizations through stages of change and success. You may write such a plan as an independent entrepreneur. Within established organizations, a team of managers, directors, and sometimes executive board members formulates the plan. In general, business plans fall into six basic types:

- Start-up plan
- Operations plan
- Internal plan
- Strategic plan
- Feasibility plan
- Growth plan.

Start-Up Plan

If you have an idea for a new enterprise, based perhaps on an innovative product or service, you need a plan to create that new enterprise. This is a start-up plan. It

DOI: 10.4324/9781003093763-18

presents evidence about the viability of your enterprise addressed mainly to potential investors. You include some or all of the following kinds of information: vision and mission statement, market analysis, sample sales and expense forecast, and other specific information that help make the case for a viable business development. In a start-up plan, you should also include

- Staffing and hiring strategies
- Leadership and resource management
- Guidelines for growing or sustaining the proposed business through human resources.

The key to a good start-up plan is to include a clear business objective and how you plan to achieve your goals. Imagine yourself as a contestant on the ABC show *Shark Tank*—how might you convince a panel of experts that your business idea is relevant to the current market, viable, and profitable? Your start-up plan should provide all the information that convinces the reader that your business idea is worth pursuing (refer to the student example in Figure 14.1B). Include budgetary information and team qualifications in more elaborate start-up plans.

Operations Plan

Often accompanying a start-up plan is an operations plan. Whether in a start-up or established company, you may be asked to participate in the creation of an operations plan for your organization. Use an operations plan to provide a clear picture of how a team, department, or division will contribute to the overall success of the organization. Consider the human resources available in your organization and how to best maximize their potential. You can propose the allocation of resources, development of key performance indicators (KPIs), and design of communication flow within the organization. You can also map out the roles and responsibilities of individuals in the organization.

Sometimes, as when an organization is undergoing a restructuring, an operations plan is needed to re-envision the roles and responsibilities of the members of the organization. For example, if your company was acquired by a larger company, you might be asked to join the managers and directors of the new company to create an operations plan that integrated existing talents and resources to optimize how things work and how employees communicate.

Internal Plan

Most students who graduate and enter the workplace land on jobs in well-established companies. You may find yourself in a position within a company that needs an internal plan to guide its next 5–10 years of business. As companies mature, they require plans for validating prudent investments and revenue growth. When you have a new product or service idea, or recommendations to improve workflow, convey your ideas through an internal plan. The internal business plan demonstrates the readiness of your ideas, and the methodologies for accomplishing your goals. It is also a tool for sharing your ideas quickly and succinctly throughout the organization. In other words, an internal plan is both an informational packet and a tool for disseminating information.

www.udel.edu/henhatch

14.1A
University of
Delaware's
"Hen Hatch"
Start-Up Funding
Competition

Source: Courtesy of
Horn Entrepreneurship,
University of Delaware

Hen Hatch is the University of Delaware's premier startup funding competition. It provides students, alumni, faculty and staff with the opportunity to make connections and collect feedback on their ideas while competing for startup funding awards and in-kind business services totaling over $100,000.

TWO COMPETITIVE TRACKS

Hen Hatch offers two competitive tracks: (1) a student track and (2) an alumni, faculty and staff track. The student track is open to all undergraduate and graduate students who are enrolled at least half time at the University of Delaware. Teams competing in the student track may include non-students; however, only student members will be allowed to present during the competition. The alumni, faculty and staff track is open to all alumni, current or retired faculty and current or retired staff. Teams competing in this track may include members who have no connection to the University; however, only alumni, faculty and staff will be allowed to present during the competition.

THREE ROUNDS TO DETERMINE AWARD WINNERS

Preliminary round   Participants submit a 3-5 page business concept for evaluation by seasoned entrepreneurs, investors and business leaders.

Semifinal round   Participants who are judged as having submitted one of the 10 most promising business concepts to the student track and one of the 10 most promising business concepts to the alumni, faculty and staff track present a 7-minute pitch and answer questions from entrepreneurs and investors.

Final round   Participants who are judged as having submitted one of the 3 most promising business concepts in each track present a 7-minute pitch and answer questions from entrepreneurs and investors in front of a live audience.

Schedule for Hen Hatch 2017

January 16
submission deadline;
www.udel.edu/henhatch

February 13
notification of semifinalists

March 10
student semifinals

March 17
alumni, faculty and staff semifinals

March 20 - April 21
finalist mentoring

April 25
finals

GUIDELINES FOR BUSINESS CONCEPT SUBMISSIONS AND PRESENTATIONS

The best preliminary round business concept submissions will tell a compelling story that provides:

1. A profile of the typical customer
2. A concise statement of the customer's problem(s) or unmet need(s)
3. A description of the product or service you've developed to solve the customer's problem(s)
4. An answer to the question of "why will customers choose your solution over competitors (i.e., what is your unique value proposition)?"
5. An understanding of the business's basic economics (product selling price, unit cost and your basic revenue model)
6. An overview of your team's relevant experiences, skills and resources
7. Any evidence you've collected to validate your claims about the problem, customer, solution, unique value proposition, and revenue model

For the semifinal and final rounds, you will also need to enhance your story by addressing the level of market traction you've gained to date, the size of the market, how you will get your product to market (channels), what costs will be incurred in so doing, one-year and three to five year financial projections, how much money you're seeking from investors, and how you plan to spend any investor funds being sought.

Title: QuadCrew Adaptive Rowing System
Team: Sam Glandon (lead), UD Senior, Mechanical Engineering
Rishay Nathoo, Kevin Kern, & Nick Sunkler, UD Seniors, Mechanical Engineering
Wyatt Grant, UD Senior, Men's Rowing Team & Mechanical Engineering
Dr. Jenni Buckley (faculty advisor), UD Mechanical Engineering
Kevin Gruber, UD Women's Rowing Team Head Coach

1. Customer Profile

Establish the most distinctive feature or selling point of your product right at the start

The QuadCrew Rowing System (QuadCrew) will initially target adaptive rowers end-users, defined as individuals with mental or physical disabilities who engage in Olympic-style rowing for competition, fitness, and/or recreation. The typical QuadCrew customer is an adaptive rowing club or community club with adaptive members, supported primarily by monthly membership fees as well as small foundation grants. These clubs purchase and maintain an array of rowing shells (boats) for their members, including 1-person (single), 2-person (double or pair), 4-person (quad), or 8-person boats. The QuadCrew system can be used with any existing racing shell to make it usable for an adaptive rower, with one complete system required per adaptive rower in the shell.

The QuadCrew system will be particularly appealing for adaptive rowers at the more severe end of the physical or mental disability spectrum. Existing adaptive rowing equipment is extremely primitive and consists entirely of seat harnesses and boat stabilizing pontoons. This technology does not eliminate the need for an individual to have substantial upper body coordination to control the oars and contribute to the propulsion of the shell (see Section 2). Adaptive rowers with severe physical disabilities, e.g., quadriplegia or hemiparalysis, are presently unserved or underserved by existing adaptive equipment for on-water rowing and can only safely participate in adaptive erging (using an ergometer, which is an indoor rowing trainer). QuadCrew would allow them to directly apply their adaptive erging practice to on-water rowing.

Don't forget your secondary audience

The QuadCrew system may also have a secondary market for recreational boaters. At present, the recreational market consists of canoes and kayaks, which are notoriously unstable, and paddle boats that are designed and priced for rental at placid community ponds. A QuadCrew recreational user would be a head-of-household who is interested in having one human-powered watercraft capable of safe navigation on both ponds and rivers. The QuadCrew system would be marketed to this user as an add-on to a canoe or kayak that enhances the stability of the water craft over oar propulsion and also allows for full-body workout. Like the target adaptive users, first adopters in the recreational boater market may also have some overlap with healthy individuals who wish to take their indoor erging practice onto the water.

2. Unmet Need

State the market niche for your proposed product

The QuadCrew Rowing System will allow individuals of all ability and experience levels to safely propel a range of recreational and competitive sporting watercraft. The beachhead market will be community-based Olympic-style rowing clubs with adaptive (disabled) rowers that are unserved or underserved by existing adaptive equipment. A potential secondary market would be recreational boaters who wish to have a single safe and versatile watercraft.

14.1B
Segments 1–4 of a Start-Up Plan in Response to the "Hen Hatch" Competition
Source: Courtesy Jenni M. Buckley

3. Product Description

The QuadCrew Rowing System is a propulsion system that can be mounted quickly and easily to any open-hulled personal watercraft (Figure 1). This includes all racing shells, from singles to 8-person shells, canoes, small skiffs, rowboats, and some kayaks. The system consists of a pull-cord that sits midline on the boat, that connects to a geared series of shafts that ultimately drive two small paddlewheels, one on each side of the boat. The paddlewheels automatically settle at the appropriate height in the water via universal joints attached to the driveshaft as well as flotations mounted to the outside of the wheels. To propel the boat, the user sits in their usual position on the boat and pulls the pull-cord back towards themselves.

Figure 1. Computer aided design (CAD) rendering of the Quadcrew design on typical rowing shell.

The propulsion system's gearing and novel self-recoil mechanism were engineered to work and feel to the user exactly like an ergometer (Figure 2). Using proper ergometer form, we estimate that the user can achieve approximately 38% of the propulsion of a competitive rowing, oar-driven system. However, proper form is not necessary to accomplish propulsion, and we will encourage our adaptive rowers to use the same modified technique that they use for adaptive erging.

Figure 2. Rowers using (left) the first-generation QuadCrew prototype and (right) a typical ergometer. The form is identical on both systems.

14.1B
(Continued)

A first-generation prototype of the QuadCrew Rowing System has already been fabricated and used in design validation (Figure 3). The prototype was designed and built by members of the Hen Hatch team as part of Senior Engineering Design (MEEG401/BMEG450). End-user feedback was solicited by former competitive rowers and current coaches during validation testing in the water tanks at the UD Boathouse. Feedback was very positive, including fast set-up times, closely matching the feel of an ergometer, and willingness to use the system on actual watercraft. Next steps are to overhaul the current prototype into a more streamlined system that fits onto a 2-person racing shell. This overhaul will be accomplished by Team Lead, Mr. Sam Glandon, as independent study over the Winter and Spring 2017 academic terms. We will perform on-water testing to validate boat propulsion and end-user ergonomics by the middle of the Spring 2017 term. This prototype will also be utilized for design pitches and demonstrations to potential manufacturing and business partners.

Figure 3. The first-generation prototype of the QuadCrew system, developed in 2016 Senior Engineering Design, by members of the Hen Hatch team. This system was used for validation and proof-of-concept in controlled, water tank testing.

Demonstrate your knowledge of the market competition

4. Unique Value Proposition

The QuadCrew Rowing System is completely novel within the adaptive rowing equipment market. The market leader in adaptive rowing equipment is Wintech (Bridgeport, CT), which is a major manufacturer of racing shells and peripheral equipment, e.g., seats, rigging, oars. Their adaptive rowing portfolio includes so-called “fixed seats” (seats that do not slide within the boat) with 2 or 4-point harness systems to support the adaptive athlete. They also sell pontoons that mount under the shell’s rigging and prevent capsizing. These equipment cannot fully accommodate adaptive athletes with less-than-full control of their arms and hands. From our site visits and end-user interviews, we have found that these individuals either (a) do not participate “on-water” – that is, they use the ergometer exclusively; or (b) go “on-water” in a 2-person shell with an able-bodied rower essentially towing them. In the latter case, end-users have expressed a desire to contribute more effectively to the propulsion of the boat.

Our QuadCrew system would be the only solution on the market that would allow severely disabled athletes to effectively row “on-water.” It would be a market expansion, rather than competition, for the market leader, WinTech, and would thus be of interest as an addition to their adaptive equipment portfolio. We have been in contact with WinTech about the concept, and their adaptive equipment specialist has expressed interest in seeing an on-water demonstration of the system. This individual also put us in contact with USA Paralympic Rowing, which also wanted to discuss our concept as soon as we could demonstrate its use on a racing shell. Again, we anticipate being able to do this by middle of Spring 2017.

14.1B
(Continued)

As the audience for an internal plan would be employees of a business organization, consider using insider language and references that appeal to the reader in a more direct way.

Strategic Plan

Sometimes, organizations may experience issues that interfere with their business or operations. Such issues may be caused by economic changes, natural or human-made disasters, social/cultural crises, emerging trends and technologies, and competitions from new products or services. For example, if you were in a business that was located in southern California, an area severely harmed by the 2020 California wildfires, you might be charged to help create a plan to recover from damage or loss.

Business organizations need a strategic plan to define the roots of problems and outline ways to address them. As a technical professional, you can help create the strategic plan. The key word in a strategic plan is *strategy*. So, include a comprehensive set of steps for solving the existing problems and prevent them from happening in the future. Your strategies should be realistic and actionable. Conduct critical analyses of the problems and propose reasonable solutions with their costs and benefits in the strategic plan.

Feasibility Plan

While a business plan seeks to explain to its readers how something might work, it also borrows from features of a proposal, which persuades readers to take action or provide support owing to a particular rationale and the feasibility of a proposed plan. Owing to changing technologies and evolving workplace practices, new procedures, workflows, and products are often needed to keep companies apace with current trends. As a technical professional, you may consider new ways to work and address customer needs. Your role in a business company may involve creating a business plan that calls attention to a pressing topic and making justifications for a specific plan of action. This plan is commonly known as a feasibility plan.

The goal of a feasibility plan is to determine if an idea, invention, method, procedure, or program is feasible. It can serve as a proposal for a feasibility study or a report of its result. In your feasibility plan, include a preliminary analysis of needs to help determine achievable goals. Then, as for a research proposal, describe the guiding questions for the feasibility study (e.g., "Is my company prepared to launch a fundraising campaign?" "What does my company need in order to launch a fundraising campaign?"). Then, describe your methods for studying your organization and the ways to report your findings. The "Executive Summary of a Feasibility Plan" box shows a sample feasibility plan consisting of these components.

Risk evaluation is a key activity in the putting together of a feasibility plan. While you should focus on the major risks that will affect your business, consider the following kinds of risk:

- Operational/human resources: Who are the right people for the job? What would happen if you couldn't hire the right personnel?
- Financial: Will your venture be affected by certain economic conditions? What are the projected interest and exchange rates?

- Technological: What if tech fails? What are the time and cost for developing necessary tools? How can you ensure trademarks, patents, or copyrights?
- Regulatory/political: How can you ensure legal compliance? Will there be any political effects caused by your venture?

EXECUTIVE SUMMARY OF A FEASIBILITY PLAN

In July 2022, The Deutsch Group, Inc., evaluated the Smithson site for federal and state jurisdictional watercourses and wetlands for the LRC Development Corporation.

The Smithson site is located in an area zoned as mixed commercial and residential, approximately 4 miles northwest of the City of Wilmington, in New Castle County, Delaware. LRC Development Corporation plans to develop this 22.5± acre site as an assisted health care community.

The purpose of this evaluation was as follows: (1) to delineate the U.S. Army Corps of Engineers' (USACOE) area of jurisdiction, which includes all areas identified as "waters of the United States," including wetlands; (2) to identify jurisdictional areas regulated by the State of Delaware as subaqueous lands; and (3) to provide documentation to support the location of the delineated wetland boundaries.

The evaluation indicates that three small tributaries and 1.8 acres of wetland exist on the Smithson site. The wetland areas are associated with the tributaries that traverse the southwest corner of the site. The wetland areas identified are characteristic of a palustrine forested wetland area.

Any proposed development of the site should attempt to avoid and, if not possible to avoid, then minimize the impact on these identified watercourses and wetlands. A Federal 404 Permit will be required for any proposed impacts on watercourses and wetlands. A State Subaqueous Lands Permit will be required only for any proposed impacts on watercourses.

Growth Plan

Inspired by the notion of continuous development in design thinking, many companies adopt a growth mindset for their business model. As your organization seeks ways to enhance existing services or expand your reach, you need a growth plan to guide your decision-making. You may participate in the creation of a growth plan for your company whether you are a designer, analyst, project manager, or director in the organization.

A growth plan can be both internal and external facing. In it, you should identify the current strengths, weaknesses, opportunities, and threats (SWOT) of your organization, segments of clients or stakeholders, objectives for growth, value proposition (i.e., why others should pay attention to your organization), and tactics for making a difference in existing practices.

- Strengths: What is your business good at?
- Weaknesses: What are you not good at?

- Opportunities: What possibilities do you see for your organization?
- Threats: What is on the horizon that may potentially harm your organization?

A growth plan needs to be adaptable to changing conditions (e.g., turnover of staff, economic development). You should propose a sustainable mechanism for short-term (5 years) and long-term (10 years and beyond) growth.

IDEATING WITH A JOURNEY MAP

It's no small feat to come up with innovative ideas that solve complex business problems. However, such efforts may begin with a design-thinking method called journey mapping—a visualization exercise to examine how certain users or consumers experience a product or service. This method allows business plan writers to understand the needs and requirements of stakeholders by walking in their shoes.

A journey map represents the process involving a series of actions mapped onto a timeline, coupled with the thoughts and emotions of an involved stakeholder to create a narrative around a product or service.

This visualization method provides just enough details for business plan writers to craft a story around a product experience and present potential ways to improve that experience.

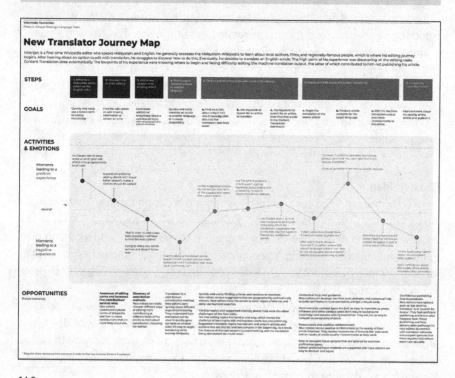

14.2
Customer Journey Map
The journey map shows the experience of a new Wikipedia editor as they translate content across wikis and use the Content Translation tool

Source: Wikimedia Commons. Retrieved from https://bit.ly/2RS7GJm

THE STRUCTURE OF A BUSINESS PLAN

A solid business plan should be a "game plan" for a successful business. A business plan contains some elements similar to those found in a proposal. They also have unique elements that address business needs:

- *Executive summary*: similar to a proposal, provides a quick glance of the business plan that highlights the central issues and recommendations.
- *Overview and objectives*: identifies the opportunities, areas of growth, or problems to be addressed. Articulates the goals of the business plan and the expected outcomes of the proposed direction.
- *Products and services*: provides an inventory of current or aspirational products and services the business offers. Compares your products and services with those of your competitors. Summarizes any needs you have in order to create improved product or service delivery.
- *Market opportunities*: analyzes and summarizes consumer demographics, habits, trends, and openness to new innovations. Creates a niche for specialized products and services that may distinguish your business from competitors.
- *Sales and marketing*: describes how you plan to create and maximize exposure to sell your products or services. Includes unique value propositions, pricing strategy, distribution plan, and marketing activities. Explains how your consumers might experience your products or services.
- *Competitive analysis*: presents a comprehensive analysis of similar products or services on the market and potential competitors. Shows strengths and weaknesses of select competitors—their sales and marketing strategies, product and service offerings, and competitive advantage. Distinguishes your business from these competitors.
- *Operations*: explains your organization's human resources, workflow, and communication practices. Showcases the specialty of your business in terms of talents as well as facilities or available equipment.
- *Management team*: provides details about the qualifications, experience, and expertise that your leadership team brings to the organization. Identifies key players and the influence they cast on the organization, in terms of managing resources, recognizing opportunities, addressing challenges, and promoting productivity.
- *Financial analysis*: fills in the numbers and analytics to tell your readers how your business is viable over time. Uses data-driven evidence to generate an accurate sales forecast. Makes the case for a profitable outlook for your business and convinces potential investors or venture capitalists to back your organization.

WRITING A LEAN BUSINESS PLAN

While business plans need to be comprehensive, they also need to be brief. To write briefly, consider employing the "lean" format recommended by the U.S. Small Business Administration (SBA). It is useful for small start-ups or companies that plan to regularly refine their business plans. The SBA recommends the format developed by Alex Osterwalder (see Osterwalder & Euchner, 2019) called the *Business Model Canvas*.

The Business Model Canvas focuses on nine components (see box) to describe your company's value proposition, infrastructure, customer/market projections, and finances.

NINE COMPONENTS OF THE BUSINESS MODEL CANVAS

- *Key partnerships*: Note the other businesses or services you'll work with to run your business. Think about suppliers, manufacturers, subcontractors, and similar strategic partners.
- *Key activities*: List the ways your business will gain a competitive advantage. Highlight things such as selling directly to consumers, or using technology to tap into the sharing economy.
- *Key resources*: List any resource you'll leverage to create value for your customer. Your most important assets could include staff, capital, or intellectual property. Don't forget to leverage business resources that might be available to women, veterans, Native Americans, and HUBZone businesses.
- *Value proposition*: Make a clear and compelling statement about the unique value your company brings to the market.
- *Customer relationships*: Describe how customers will interact with your business. Is it automated or personal? In person or online? Think through the customer experience from start to finish.
- *Customer segments*: Be specific when you name your target market. Your business won't be for everybody, so it's important to have a clear sense of who your business will serve.
- *Channels*: List the most important ways you'll talk to your customers. Most businesses use a mix of channels and optimize them over time.
- *Cost structure*: Will your company focus on reducing cost or maximizing value? Define your strategy, then list the most significant costs you'll face pursuing it.
- *Revenue streams*: Explain how your company will actually make money. Some examples are direct sales, memberships fees, and selling advertising space. If your company has multiple revenue streams, list them all.

Source: Small Business Administration. Retrieved from https://bit.ly/3xcNvpB

DELIVERING THE BUSINESS PLAN

The way your business plan looks says a lot about its worth. Review your draft for style and visual design.

Style and Visual Presentation

While most business plans are presented orally before a panel of investors, print-based plans are necessary for reference. Design your business plans to be visually appealing and engaging.

For more about oral delivery strategies, see Chapter 19

For more about data visualization, see Chapter 9

Consider the style of your design. What impression do you want to convey? Exciting? Intriguing? Serious? Creative? The answer is likely a combination of these emotions. Make explicit choices as you decide on the design of your document, including such visual elements as colors, arrangement and signaling of segments, and typography. You can help your audience access complex data quickly by using graphs, charts, tables, and infographics.

Be sure to adhere to your business's or company's brand authority by using its existing identity communication guidelines such as logo use and placement, font selection, photography, brand voice and tone, and other visual policies. Figure 14.3 provides examples of visual elements you should pay attention to when designing a business plan. Figures 14.4 and 14.5 show a sample brand identity manual for an organization to specify visual guidelines for designing official documents.

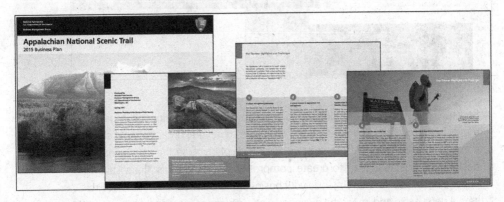

14.3

Visual Elements of a Business Plan

Source: Appalachian National Scenic Trail—Business Plan (2015). Retrieved from https://bit.ly/3gsYNAa

---

1.5   INTRODUCTION

Terminology

Logo elements

- "St. C" Logo – The "St. C" is St. Cloud State University's primary logo.
- Wordmark – St. Cloud State University's graphic name identity, which appears in stacked and horizontal applications.
- Tagline – Education for Life is the university's tagline. Unlike a campaign, the tagline has a long life that is clear, authentic and straightforward – critical components of the brand strategy.
- Unit signature – The name of units within St. Cloud State, such as colleges, schools or departments included in a unit logotype.
- Logotype – The combination of the "St. C," wordmark and other elements including the tagline and unit signatures.

Campaign slogan – A campaign slogan differs from a tagline. It generally has a shorter life and is tied to marketing where the tagline is tied to the university identity. i.e. "Think. Do. Make a Difference." is a university campaign slogan. "Education for Life" is the university's tagline.

Clear space – Minimum amount of space required to separate the logos and logotypes from other elements.

Color models

- RGB – Color model in which red, green, and blue light are added together to reproduce a broad array of colors. Primarily for electronic display.
- Pantone – Pantone Matching System (PMS) is a proprietary color space used in a variety of industries, primarily printing, though sometimes in the manufacture of colored paint, fabric, and plastics.

Ghosted 'St. C' and Huskies (secondary) logo One-color versions of the "St. C" and Huskies secondary logos with screened or multiplied opacity that is used as a translucent graphic element over flat or patterned backgrounds.

Trademark

Trademark identification ("TM") is a required element of the "St. C" and secondary Huskies logos. If an exception is needed, contact University Communications. Trademark is not required when using the wordmarks or taglines alone.

14.4

Guidelines for Use of Terminology in Official Publications

Business plans help organizations to foster innovation and continuous improvement. In this chapter, you have been introduced to several organizational situations for business plans, from start-up to growth plans. Use the strategies for designing a business plan to create comprehensive as well as "lean" plans, depending on your needs. Organize your business plan with a clear structure. Present a persuasive and impactful business plan with appealing visual treatments.

CHECKLIST: BUSINESS PLANS

1. Determine the appropriate business plan for your organizational situation

    Start-up plan: states a business opportunity and other specific information that make the case for a viable business development

    Operation plan: describes the operations design and management of your business

    Internal plan: validates prudent investments and revenue growth

    Strategic plan: defines the problems that affect business operations and outlines ways to address them

    Feasibility plan: explains how something might work and persuades readers to take action or provide support

    Growth plan: identifies the current strengths, weaknesses, opportunities, and threats of the organization; segments of clients/stakeholders; objectives for growth; value proposition; and tactics for change

2. Design your business plan to be a "game plan" for success

    Include executive summary and objectives/overview sections to articulate the business plan's goal

    Provide an inventory of current product/service and analyze market opportunities

Explain your sales and marketing strategies

Analyze existing market competition

Describe your company operations, including operation and leadership teams, and financial well-being

3. Create a lean business plan to help make decisions efficiently
4. Design your business plan to be visually engaging and persuasive

Use data visualization (e.g., charts, tables) to communicate complex information

Choose visual styles that reflect the nature of your plan

Adhere to your organization's brand identity

EXERCISES

1. Examine the Deutsch Group feasibility plan shown in the "Executive Summary of a Feasibility Plan" box. How has the plan addressed the following kinds of risk?

   - Operational/human resources
   - Financial
   - Technological
   - Regulatory/political.

2. Create a journey map for your university library experience. How does a student experience the process of browsing the catalog, locating a book, and checking it out? Identify the ups (positive experience) and downs (negative experience) of that process.

FOR COLLABORATION

Collaborate with three or four classmates to create a basic start-up business plan using the University of Delaware Hen Hatch start-up funding competition submission criteria. Read its guidelines for business concept submission carefully and begin with the following questions:

- What is your proposed business concept? What needs does your product serve?
- Who is on your team, and what are the roles and qualifications of the team members?
- How did you decide on the design of your business concept?
- Why is it important to fund your business concept?
- What are the basic economics for your business concept?

Use the student example in Figure 14.1B as a model.

REFERENCE

Osterwalder, A., & Euchner, J. (2019). Business model innovation: An interview with Alex Osterwalder. *Research-Technology Management, 62*(4), 12–18. Retrieved from https://doi.org/10.1080/08956308.2019.1613114

15 Brief Reports

Often, you will produce brief reports for different situations in the workplace. These short reports, as opposed to full, comprehensive reports, are common artifacts for informational reporting and record keeping. In this chapter, you will learn about different types of brief reports, their features, and how to compose and deliver them effectively.

PURPOSE

Brief reports convey interim information that gives your colleagues or supervisors an update about the status of your work. Although brief reports may seem unnecessary or to interrupt your project workflow, writing these reports helps you:

- Check that you're keeping on schedule
- Record work accomplished over a specified period
- Convince a client, sponsor, or adviser that your work is on track and on schedule
- Assure readers that there will be no surprises at the project's end.

The first two purposes serve you directly. The interruption each report offers lets you survey and appraise the project as a whole and evaluate one phase. The report deadline may also motivate you to finish the work, so you avoid embarrassment before your supervisor or your peers. In addition, writing during the project helps you remember details and, if the project works, approach the task of writing the final report with some text already at hand.

The third and fourth purposes pertain more to management. Supervisors and administrators consider progress reports essential managerial links. Such reports foster intelligent decisions on matters of money, time, equipment, and trends that may show whether a project should be continued, reoriented, or abandoned. Submitted to a sponsor, these reports often inspire renewed confidence in the feasibility of the work, help capture any necessary additional resources, and convey an optimistic image of you and your organization.

SITUATIONS

Although the most common brief report you'd be called on to produce in the workplace is the progress report, there are other types of brief reports that serve to address specific situations (i.e., meeting minutes, periodic activity reports, and incident reports).

DOI: 10.4324/9781003093763-19

These reports document discussions that took place during a meeting, activities that occurred during a period of time, and incidental events. In this section, you will learn about the components of these reports and how to compose them.

Meeting Minutes

You will attend meetings when working on projects. It is good practice to document meeting agendas, proceedings, and discussion outcomes. The term "minutes" has been considered a derivation from the Latin phrase *minuta scriptura*, which means small or rough notes. Meeting minutes record who attended a meeting, what was discussed, and any deliberations and decisions made. They serve as a reminder for you and your colleagues about everyone's responsibilities. Minutes are typically recorded by an administrative assistant or an appointed individual, but it is becoming increasingly common for minutes to be written collaboratively. Using a web- or cloud-based word processor such as Google Docs or Microsoft Word via OneDrive, team members can open a shared document and document meeting notes synchronously.

For more about writing collaboratively, see Chapter 4

Key components of meeting minutes include the following:

- Meeting title, date, time, and location
- Attendees and/or participants (including guests, observers)
- Agenda items (pre-arranged as well as additional topics that emerged during the meeting)
- Decisions (if any; include deliberation and voting process, results)
- Future topics or unresolved issues.

MEETING MINUTES TEMPLATE

Meeting Name
Date:
Time:
Location: [Physical or virtual venue]
Attendees: [List all names]

Agenda Items
1. [Insert item]
2. [Insert item]

   - [Use sublevel bullets to add descriptions]
   - [Use sublevel bullets to add descriptions]

3. [Insert item]
4. [Insert item]
5. [Insert item]

Action Items Owners' Deadline Status
[Item 1] [Name(s)] [Date] [In progress/completed]
[Item 2] [Name(s)] [Date] [In progress/completed]
[Item 3] [Name(s)] [Date] [In progress/completed]

Meeting minutes should be completed and shared with the team—and stakeholders, if applicable—immediately after the meeting. Minutes are typically emailed to these relevant recipients. They can also be posted to a website for open access if the project is federally or publicly funded.

Periodic Activity Reports

When working on long-term projects, you will need to provide regular updates to your supervisors or managers. Periodic activity reports provide quick "status updates" for a project and resemble a progress report. Unlike progress reports, which are summative and more comprehensive, activity reports focus on a particular event or activities that occurred within a small time frame (daily or weekly reports). In a periodic activity report, you should describe key accomplishments in an ongoing project and any noteworthy developments.

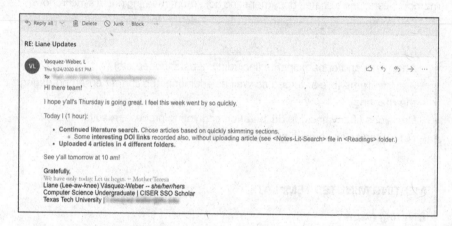

15.1
Activity Report
Delivered via Email
Source: Courtesy of
Liane Vásquez-Weber

Incident Reports

Another type of activity report you may compose is the incident report. You use incident reports to document something that has happened, usually unexpectedly. An incident report records details of a workplace accident or event that requires managerial attention. The purpose of this report is to account for the circumstances or conditions that led to the incident, factual descriptions about the incident (what happened, when it happened, who was involved, what was the impact, etc.), and viable corrective actions to remedy any problems caused by the incident.

The main goal of both periodic activity and incident reports is record keeping. Therefore, it is important to ensure accuracy in the reporting of events or accomplishments, and timeliness in the completing of these reports for administrative processing.

INCIDENT REPORT

Arrival
Night Security Staff AHMED, Charith called Community Advisor SMITH, Fiona at 1:32 a.m. regarding an incident where Resident LOTEZ, Shoko from Laurie Hall 2nd Floor has injured her thumb by slamming the car door on her thumb in front of Laurie Hall.

Observation
LOTEZ was brought to sit in front of the Laurie Hall front desk. By the time SMITH arrived, two (2) Night Security staffs, SUBA, Amresh and AHMED, two (2) Public Safety staffs, DEMKO (#309) and WALZ (#208), were present.

SMITH observed that LOTEZ has a minor injury on her thumb and it has stopped bleeding. SMITH observed that LOTEZ may have been drunk. She looked tired and had dilated pupils.

Public Safety staff reported that St. Cloud Police has been informed about this case and they will be arriving anytime to check LOTEZ's situation.

Statement by Takemoto
"I slammed the door on my finger . . . Am I going to get into trouble? I drank some beer . . . I'm a minor." LOTEZ showed worries and she was physically tired.

Action
Two (2) policemen came at 2:45 a.m. and observed LOTEZ. They checked on her thumb and checked her breath. Because this was a case of minor drinking, the policemen issued a ticket to LOTEZ. She was asked to wait for a mail and to follow the instructions on the letter. This incident was filed as case no. 44372, according to the policemen.

Conclusion
The policemen left at 3 a.m. and LOTEZ was sent to her dorm room.

SMITH reported the incident to Area A duty person, CILANTRO, Danny.

Progress Reports
When working on larger projects, you are required to keep your funders and managers posted about the status of your work. You will create comprehensive progress reports—not just periodic update emails—to communicate your project status.

Progress reports contain more substantive information about an ongoing project than an activity report. A standard approach to reporting progress includes:

For more about funded projects, see Chapter 13

- Work completed
- Future work
- Problems (if any)
- Correction/solution (if there's a problem)
- Schedule status
- Budget status (if applicable).

A full-in-the-black approach, however, will not work in every situation. More complex projects may require more details on their progress. The core categories remain the same, but you need to provide explanations that serve your persuasive purposes. You may also have to amplify the core content with other segments, such as an introduction, a conclusion, or recommendations.

Introduction

Introduce the core information with an opening statement that provides that context for the report. State that the document is a progress report, note the period covered, and summarize briefly, sometimes in quantitative terms, the status of the project. You may also, especially in a lengthy project, review the problem and objectives. Here's how a team of engineering students introduced their report to the faculty adviser:

> This progress report covers CS Environmental's activities between 1 April and 5 May on our senior design project. The goal of the project is to evaluate geological and soil conditions at the site of the proposed office campus and determine that best foundation type for Buildings A and B and the steel tanks. As of 5 May, the project is 50 percent complete, as scheduled.

Another student team began its report on the redesign of workstations at a telemarketing firm with a summary of past and future work:

> The following report details our progress on the redesign of your workstations. We have completed the first phase outlined in our proposal of 12 October. In completing the phase, we have determined the problem in the current workstations that cause employee discomfort and contribute to your rising health care costs. We will spend the remainder of the project concentrating on these areas. Our next steps are to survey equipment currently available on the market that can address these problems and determine if installing this equipment would be cost-effective. We will also develop guidelines for reeducating employees to protect themselves. Where possible, we will suggest such reeducation as an alternative to new equipment. The rest of this report briefly summarizes our findings to date.
>
> (Courtesy of Mary Angerer and Rebecca Tinsman)

Work Accomplished

In the middle of the report, you assess progress toward your goal. The following discussion covers the most common approaches to presenting this information.

- *Chronological*: The easiest way to present your work is chronologically: what you did from Day 1 until now, as in this example:

  1. On December 1, I mailed 450 questionnaires to recreation directors in 20 Texas cities.
  2. On December 6, I interviewed the mayor and the director of Austin Recreation. Both approved of the program, although they questioned the feasibility of financing.
  3. On December 8, I obtained a copy of Houston's program and found it similar to that in Austin.

  Such a chronological approach, however, is usually not the best one. It makes the reader work too hard, and this may not be convincing. It lacks emphasis and may relegate to a subordinate position some key event—because it occurred, for example, on a Tuesday 5 weeks into the project—that should be highlighted to aid in a decision on the future course of the project.
- *Task by task*: You may also plan the progress report to reflect the tasks outlined in the proposal. Detail the status of work on each task and the extent to which they have been completed according to the proposed directions.
- *Component by component*: To report progress on a mechanism or system, proceed from one component to another component. For instance, the student report on the redesign of workstations divides its description of the work accomplished into the components of their redesign: video display terminal, keyboard, and chair. For each component, the team first analyzed the symptoms that appeared as a result of poor design and the features of the current stations that contribute to these problems. They then recorded the results of that analysis as a baseline for the redesign.
- *Decision support*: Organizing information by time, task, or component helps you describe your work. Sometimes, however, you need a structure that's more overtly persuasive. You build an argument that aids the reader in making a decision about the project and assures the client or sponsor that all is under control. If, for example, you have encountered a problem, show how you plan to overcome it:

  Although we had anticipated including 22 buildings on the tour, we discovered after a preliminary run through the route that we could not complete the tour in the time allotted. Traffic delays—the normal rush hour chaos in Center City—are up 1 half-hour. It took us an average of 8 minutes to herd people on and off the bus, another time factor we had not calculated in our preliminary design. The mechanics of simply getting from place to place reduced the impact and enjoyment of the places we did visit. We have thus pared the list of sites to 10, a number that allows us to sample the breadth of architecture in the city without sacrificing the ability for participants to savor what we present.

Future Work

At the end of the report, you generally provide a look ahead to work in the next phase. State the prognosis for the project in a tone that avoids either unbridled optimism or dark despair. If the project is not working—a condition you should make clear from the outset of the report—the closing section recommends a change of tack or scope, maybe even abandonment if the problem looks unsolvable. But, when the evidence warrants, the closing provides assurance of timely and successful completion. End on a positive, confident note. Here is a brief closing in list form:

The tasks remaining include:

1. Finding information about construction time for precast concrete structures
2. Deciding on a framing system based on that information
3. Consolidating individual team member reports and editing them into a final report that provides our recommendation.

Overall, this project has gone smoothly with few complications. The team has worked together effectively, and we are confident that you will be pleased with the final report.

Table 15.1 Simple Schedule					
Task	**Wk 1**	**Wk 2**	**Wk 3**	**Wk 4**	**Wk 5**
Decide approach (all)	✓				
Visit site (all)	✓				
Interview 3 employees (RR)		✓			
Interview 3 employees (PX)		✓			
Interview 3 employees (JJ)		✓			
Review medical records (RR)			✓		
Review workstation literature (PX)			✓		
Write progress report (JJ)			✓		
Determine design options (all)				→	
Select best design or method for dealing with current work stations (all)				→	
Write final report (JJ)				→	→
Edit report (all)					→

Compare proposed work with the current status. All items marked with a ✓ have been completed; those marked with → remain for the next project phase (Weeks 4 and 5). Initials indicate the team member responsible for the task

Table 15.2
Budget Segment in a Progress Report

Item	Budgeted ($)	Actual ($)
Airfare to Minneapolis	400	325
Hotel (2 nights)	250	275
Per diem	100	100
Conference registration	300	300

The amounts in the "actual expenditure" column would be totaled (you see only a segment of the budget here) and that amount subtracted from the project budget to provide the current balance.

Schedule and Budget Status

In addition to these discussions of work done and to be done, most progress reports also include two other segments, usually in visual form: a schedule and a budget. In these segments, you show you've spent the allotted time and money and how that expenditure compares with what you proposed at the beginning of the project. Some progress reports focus almost exclusively on these two segments. The visual form helps you emphasize the comparison between projected and actual status (see Tables 15.1 and 15.2). When that comparison shows significant deviance—that is, unbudgeted expenses or delay in the schedule—explain those problems in the text.

BEST PRACTICES FOR BRIEF REPORTS

Just because brief reports are usually seen as informal reports, it does not mean these reports should be done casually. Good reports address problematic issues and keep projects on track. To accomplish these goals, it is important to ensure clarity, timeliness, and practicality when composing brief reports.

Clear Focus

Brevity and clarity are the key features of brief reports. Whether you're reporting a periodic activity or meeting discussion, you should aim to capture important information with accuracy and focus. Report readers tend to just want the bottom line. Avoid giving unnecessary details about an event or the progress of a project. Use clear subject lines, headings, bulleted lists, and short paragraphs to break up dense content and present essential information efficiently.

For more about document design, see Chapters 9 and 10

Timely Information

Good reports provide just-in-time information to the reader. Old or background information does not need to be included in a brief report unless it is crucial for context purposes. Reports such as meeting minutes and incident reports should be composed and delivered to relevant recipients quickly so they can be used for future planning or corrective actions.

Actionable Items

When reporting recommendations or indicating a future plan, be sure to present them in actionable terms (i.e., things your readers can do). This means writing in an active voice and including criteria for deliverables, such as a project timeline (what will be done, by whom, and when), completion measures (how do we know the work is done?), and quality measures (how do we know it is done well?). Your report may also anticipate questions from its readers and offer answers based on available information. In some cases, a call to action is necessary in the conclusion of a brief report to prompt reader feedback or reaction.

COMMUNICATING BRIEF REPORTS DIGITALLY

To reiterate, the main function of brief reports is to communicate small bits of information to relevant readers quickly. As noted in the previous section, timeliness is expected in these reports. For this reason, as well as matters of convenience, brief reports are frequently produced and shared digitally using technologies such as emails, instant messaging, digital forms, blogs, and social media.

Emails and Instant Messaging

Today, it's common for brief reports to be created and circulated within instant messaging communication environments such as emails and applications such as Microsoft Teams, Slack (popular in European and American workplaces), WeChat (popular in Asian workplaces), or WhatsApp. These platforms allow the composition of brief reports with multimedia supplements, including voice or audio files, images, and videos. Although emails are a standard tool for business communication, chat apps such as the ones named above are becoming regular channels for business-related exchanges. Organizational communication researchers have found that chat apps promote multitasking and increased interactions among coworkers. However, it is important to remember where the boundaries are between work and personal life. If you and your colleagues use chat apps to communicate, it would be wise to create an account for business use that is different from your personal account to separate non-work-related communication.

For more about online correspondence, see Chapter 18

Digital Forms

To save time and increase documentation efficiency, some organizations have opted to use digital or online forms to create and submit brief reports. Digital forms can be found on a company's intranet (accessible only by employees) and can be shared via URL or QR codes. These forms contain common fields and entry sections in order to maintain consistency across reports. Digital forms are usually linked to a database that is populated with the contents entered into the forms. This helps organizations save time by skipping the need to employ human labor and time to collate information from paper reports.

Blogs and Social Media

Depending on their context and audience, brief reports may be delivered via blog postings or social media channels. For example, a team may share a status update report (periodic activity report) with its members through a private Facebook Group created for a project. It is common for federally funded projects or crowdsourced activities to share progress reports with the public audience using blogs or dedicated platforms. Figure 15.2 shows a blog post example that shares new findings from a grant-funded project. Figure 15.3 shows a brief update by the creators of a crowdfunded project on Kickstarter.com. Note the different levels of informality between the two examples and their effects.

15.2
Blog Post Communicating New Findings from a Grant Project

Source: Courtesy of Vega Academic Publishing System

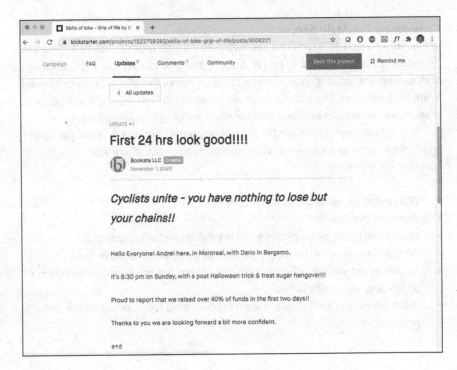

15.3
Brief Update from Creators of a Crowdfunded Project on Kickstarter

Source: Courtesy of Bookata

AUDIENCE AWARENESS

As with the process of composing any technical and professional artifacts, you should consider your tone and style in presenting brief reports to ensure desirable reactions from your audience. Establishing the appropriate persona for your reports and understanding the needs of internal and external audiences are core design-thinking practices that can help you achieve desirable outcomes and be an effective communicator.

Persona

The persona you manifest in writing depends on the purpose of your report. Again, the informal nature of brief reports does not mean they should be unserious. The persona you establish in a progress report, for instance, should be authoritative and convincing. When addressing problems or issues, or when you are proposing a budgetary update in a project (e.g., asking for more funding), you want to present a credible persona through the tone and language in your report. This helps build your readers' confidence in your recommendations.

When dealing with troublesome situations, avoid using ambiguous language that hides the problems. As one veteran reviewer of progress reports remarked, "When the contractor says they are 'virtually on schedule,' how far behind are they? It's usually about 6 months." You want to sound positive in a progress report, and, thus, some bending of the truth may be appropriate, especially because you may well be legitimately uncertain about outcomes. However, you risk behaving unethically if you pad a report with meaningless phrases that can be interpreted in a number of different ways. Humor, some experts suggest, may be suitable for such padding.

Considering Public Audiences

In Figures 15.2 and 15.3, you can see two public-facing reports that have different personas. Although both are cheery and bright, the example in Figure 15.2 balances both excitement and diligence to convey maturity (and, therefore, trustworthiness) in the report authors. The Kickstarter project creators in Figure 15.3, in comparison, present greater enthusiasm and spontaneity through colloquial writing.

These two examples illustrate the importance of context in determining the appropriate persona to present. When addressing public audiences, you should consider the following:

- Who might be reading this report?
- What are their expectations?
- What's the significance of this report? Who might be impacted by it?
- What's the nature of the report? Are these positive or negative contents?
- As the report will be publicly accessible, what are my concerns in terms of ethical and legal compliance?
- What would happen if an audience misinterpreted the intention of the report's persona?

For more about
audience analysis,
see Chapter 2

That last point of consideration can prepare you for a worst-case scenario, which may help you determine the risks versus rewards in your selection of an appropriate persona.

> Brief reports convey interim information in a timely and concise manner. You have learned in this chapter the situations requiring brief reports in the workplace. Use meeting minutes and incident reports for record-keeping purposes. Design a progress report with emphasis on the work done and/or any modifications to the proposed method. At any rate, make sure your reports are clear, timely, and actionable. For more efficient documentation, deliver your brief report digitally using online content management systems as well as social media.

CHECKLIST: BRIEF REPORTS

1. Take time to assess projects in midcourse
2. Use that assessment to monitor your work and record preliminary results
3. Meet the needs of management in monitoring projects and budgets
4. Use digital tools and platforms to compose and deliver reports, if appropriate
5. Write reports that establish the appropriate persona for particular audiences

EXERCISES

1. Send a periodic update report in the form of an email to your course instructor. Use this brief report to assess your progress in the course. Measure your current progress against your personal goals as well the instructor's learning objectives as outlined in the syllabus.
2. Assume you are the president of a student organization such as the Engineering Professionals Society, Emerging Entrepreneurs Association, or Technical Communicators Club. Prepare a progress report using the recommended structure in this chapter to report on a campus event that your organization is hosting in the upcoming month. Include programming, publicity, selection of meeting places, speakers/presenters, volunteers, and financial arrangements. Include a statement of the future work needed to complete your organization's plans.

FOR COLLABORATION

For a collaborative project in your course, set up a team communications channel using one of the recommended tools in this chapter (or find a different one on your own). With your teammates, establish some rules for best communication practices. Note the affordances and limitations of your communication channel.

With your teammates, write a collaborative progress report for your instructor. Follow the guidelines in this chapter. Include a schedule that has been updated, if necessary, to show any changes in tasks or deadlines from what you had earlier proposed. Briefly describe what you have learned so far. Note any problems you have encountered and indicate how you plan to solve them as a team.

16 Formal Reports

Many research institutes make a point of saying, "Reports are our only product." In the workplace, you will create formal reports to supplement your projects for the purposes and situations included below.

Unlike brief reports, which prioritize brevity and timeliness, formal reports are comprehensive and less time-sensitive. Formal reports help individuals and organizations find more efficient methods of production, raise profits, develop new products or markets, and meet social and environmental responsibilities. Like long proposals, formal reports often represent collaborative efforts. In this chapter you will learn how to write formal reports in different situations.

PURPOSE AND SITUATIONS

For more about explaining versus persuasion, see Chapter 8

First, determine your purpose in writing the report. That purpose also reflects the reader's reason for reading the report. Two major purposes for writing are to inform and to persuade; similarly, two major purposes for reading are to understand and to decide or act.

Although informing and persuading can overlap, it's helpful to think of one major purpose as you select and structure information. If you aim to enhance your reader's understanding of a project, an event, a site, a concept, and the like, write an information report, which is largely descriptive. If you aim to help the reader make a decision or act in a specific way, write a decision-making report, sometimes called an action report. Your approach in that context is more overtly persuasive and argumentative.

Information Report

An information report may include the results of personal observations, experiments, interviews, surveys, and meetings. It may provide background on a situation or a position. You may also write an information report to:

- Define a concept
- Describe a situation or event
- Present findings from empirical research
- Review the literature on a topic
- Show the financial status of your organization

- Show how your organization complies with environmental or other standards
- Outline a marketing approach
- Define a policy
- Summarize.

Decision-Making or Action Report

In a decision-making report, you deploy your evidence in a well-structured argument that aims to help the reader solve a problem, make a decision, or act in the way you think is right. For example, you may write a decision-making report to:

- Examine the feasibility of an action
- Justify an action by providing good reasons
- Define the options and limitations in a situation
- Provide recommendations.

An information report may lead to a decision-making report. For example, you might write an information report to summarize research comparing the reproduction rates of bald eagles in different areas of North America. Or you might describe your own empirical investigation into why bald eagles along the Maine coast are the slowest to reproduce. If you find that the birds display high levels of contamination by PCBs (polychlorinated biphenyls) and mercury, and those toxins are inhibiting reproduction, you may then write a decision-making report that recommends action to reduce pollution and increase the bald eagle's reproduction rates.

Content

Second, gather and select information. Although their purposes may differ, informational and decision-making reports can draw on the same source of information. Determining which kind of report to write, and on what topic, is usually not too difficult on the job; the intended reader often makes the assignment. In your career, you may write reports on a wide variety of topics. For example, one increasingly common topic for such reporting is the environment.

As a student learning about the construction of reports, however, you often must begin by choosing a subject to collect information about. Use your understanding of the purpose of reports to focus your choice. Let's assume, for example, that you'd like to write a report on solar energy. How can you narrow that topic? To prepare an information report, divide the topic into appropriate components or categories, as in the following list of general categories applied to solar energy:

For more about defining a problem or opportunity, see Chapter 5

- Source: wind power, tides, solar gain
- Type: active or passive
- Location: solar installations in a particular city or area
- Building: solar churches, solar residences
- Material: new materials for constructing photovoltaic cells.

To prepare a decision-making report, identify audiences who might be interested in problems or issues associated with solar energy and then gather and select information that will help solve their problem, as in these examples:

- Homeowner:
  - o Examine a house to determine the feasibility of converting its present heating system to use solar energy
  - o Design a solar system for a new house
  - o Retrofit a solar system onto an existing house.

- State energy agency:
  - o Recommend that the agency develop a program to increase awareness of solar energy among schoolchildren.

DESIGN OF FORMAL REPORTS

Third, design information to achieve your purpose. Like other genres of technical communication, reports often follow conventional approaches. Learning those approaches will help you meet your obligations to your reader and to your information. This section first teaches you how to create an abstract or executive summary to capture the essence of your report. Then, it provides guidelines for structuring report introductions, often the hardest task in writing reports. Next, it shows you how to follow up on the introduction's promises in two genres of reports: an explanatory report and an IMRAD report.

Abstract and Executive Summary

An abstract is a summary of the content in your report. Abstracts are common in academic or research settings, and another form of summary, called an executive summary, is common in business settings for corporate or government reports. Whereas abstracts rarely exceed 200–300 words, executive summaries vary considerably depending on the length of the parent document.

If your report has a preexisting structure for the abstract or executive summary, then follow the structure. If not, think of your abstract as a prose form of your outline or table of contents for the report. When you write,

- Emphasize what the report says that is new and significant
- Stress the scope and objectives of the work (the problem and purpose)
- Report the chief results, conclusions, and recommendations, if any
- Describe procedures briefly, with emphasis on any departures from the customary or original directions
- Omit supporting examples or details that can be found in the report.

In terms of voice, use active verbs whenever possible. Use past tense to report what was done, but present tense to indicate ongoing activities and implications of the activities.

Introduction

Whether you are writing an information report or decision-making report, you need to perform two critical tasks in the introduction: (1.) draw the readers' attention and (2.) orient them to the content and the structure of the report. To draw attention, consider some of the following strategies used in these model paragraphs:

- State your purpose directly:

  The purpose of this report is to present the results of a survey concerning the use of campus makerspaces during the current academic year. All 11 maker studios participated in the survey. Academic IT and Learning Resource Services (AITLRS) questioned both site monitors and student users about their level of satisfaction with the management of the studios, with currently available maker tools and networked technologies, and with logistics: hours of operation, accessibility of studios, and training programs. After a brief description of the survey methods, the report summarizes comments in each of these three categories: management, resources, and logistics.

- Establish the significance of your topic:

  The first campus makerspace mainly attracted students in technical programs such as engineering, science, and design. Recently, however, maker practices underlie instruction in every course. Many students arrive at the university with excellent general maker and design skills, especially skills at desktop publishing and exploring virtual worlds. But they often lack even a basic understanding of academic making—for example, using sophisticated modeling software and operating milling machines. Students are simply expected to demonstrate those abilities.

  Student users thus place increased demands on site studio attendants to provide instruction in these tools. Time for that instruction, however, only adds to the tasks attendants must perform and causes inefficiencies in studio usage. Moreover, the students are often frustrated in trying to accomplish their required coursework. That picture, at least, emerged from meetings with several studio attendants. To determine the level of satisfaction—or dissatisfaction—among current users of the studios, as well as attendants, AITLRS conducted the survey whose results are summarized in this report.

- Correct a misunderstanding:

  As a group of studio attendants representing each of the university's 11 makerspaces, we have compiled this report to present our concerns about the current status of the studios. The recent survey conducted by AITLRS covered some of the pertinent issues, but we feel that several other topics need to be addressed before the university will have an accurate assessment of satisfaction levels at the studios. In particular, we think the survey questionnaire did not provide enough emphasis on who is responsible for training students to use maker tools and which applications the studios should support. This report thus focuses on these two issues while also providing some additional comments on each of the three categories of information requested by AITLRS—management, shared resources, and logistics.

- Characterize your intended audience:

  This report summarizes the results of a survey AITLRS conducted over the past three weeks to determine levels of satisfaction among attendants and student users at our makerspaces. Because the report addresses an internal university audience, we have not detailed the locations and characteristics of the maker studios. The report also assumes familiarity with the tools and applications supported by AITLRS and with course structures and academic calendars at the university.

- Refer to a direct request:

  On 15 May, David Folsom, Executive Vice President, asked AITLRS to prepare a report assessing how satisfied students are with the current operation of our maker studios. To answer his question, AITLRS conducted a survey of student users. In addition, to enhance the validity of our findings, we also surveyed studio attendants. This report presents the results of both surveys.

- Bring the readers up to date:

  The last five years have seen significant changes in the maker scene on campus. The major change is a vast expansion in the population of students using makerspaces. While five years ago the population of such users was restricted largely to students in technical fields, today students are encouraged to make innovations in all their courses. These new users have intensified the demand on our current maker resources. They have, for example, increased the need for studio attendants to provide instruction on pertinent applications and hardware. Such instruction is essential, in part because professors rarely take time to tell students how to accomplish such tasks as using sophisticated design software and operating maker tools like the milling machines. Students are simply expected to demonstrate those abilities. When studio attendants have to provide such instruction, that creates inefficiencies in the use of studio resources and heightens frustration levels. Even more significantly, the burgeoning of new applications and machines has also caused pressures on AITLRS to expand its expenditures as the maker studios try to keep up to date.

These model paragraphs show how you might open a report. In addition, you might note any limitations in the scope of approach of your report. If, for example, your reader expects you to cover the costs of equipping and monitoring makerspaces, mention the reasons for that limited scope in your introduction. For most reports, it's a good idea to end the introduction with a statement about the report's place— for example, "The following report is divided into three segments, each detailing the results on one of the survey's main categories: management, resources, and logistics." That final statement then leads the reader smoothly into the report's discussion.

Explanatory Report

Each paragraph concerning campus maker studios might begin a different explanatory report about the topic. An explanatory report, as the name implies, helps the reader understand some concept, event, situation, or object. Particularly if the report centers on a mechanism, organism, system, or process, it may become an extended definition or description.

Compliance Plan

The audience may dictate the report's plan when it demonstrates compliance with a request or legal requirement. For example, consider the box that shows the "Required Structure for an Environmental Impact Statement" (EIS; one common form of explanatory report). This structure is dictated by the U.S. federal government, which requires agencies or other organizations to file reports when they propose some action, such as commercial developments, explorations for natural resources, and new parks or recreation areas, that might have potentially harmful impacts on the environment. The report begins with a description of the proposed action and then details the current environmental conditions and how they might change.

REQUIRED STRUCTURE FOR AN ENVIRONMENTAL IMPACT STATEMENT

- An Introduction including a statement of the **Purpose** and **Need of the Proposed Action**
- A description of the **Affected Environment**
- A **Range of Alternatives** to the proposed action. Alternatives are considered the "heart" of the EIS
- An **analysis** of the environmental impacts of each of the possible alternatives. This section covers topics such as:

  o   Impacts to **threatened or endangered species**
  o   **Air and water quality** impacts
  o   Impacts to **historic and cultural sites, particularly sites of significant importance to Indigenous peoples**
  o   **Social and economic impacts** to local communities, often including consideration of attributes such as impacts on the available housing stock, economic **impacts to businesses, property values, aesthetics and noise within the affected area**
  o   **Cost and schedule analyses** for each alternative, including costs and timeline to mitigate expected impacts, to determine if the proposed action can be completed at an acceptable cost and within a reasonable amount of time.

Source: Eccleston (2014)

Classification Plan

Another plan is to organize information through classification, as you can see in the following outline of a report that classifies alternative solutions to the problems of a regional transportation system into six categories:

1. Freeway alternatives:

   - Capacity expansion and operational improvements
   - HOV (high occupancy vehicle) lanes
   - Incident management
   - Ramp metering.

2. Arterial alternatives:

   - Capacity expansion
   - Operational improvements
   - Computer signal systems
   - Access management
   - Incident management.

3. Transit alternatives:

   - Rail or busways
   - Bus service.

4. Demand management alternatives:

   - Parking policies
   - Employee commute plans.

5. Land use alternatives.
6. Intelligent vehicle-highway system (IVHS) alternatives.

As the last segment of such a report, you may simply present the last topic promised in the introduction, after you remind readers briefly about that promise. The transit report, for example, ends with a discussion of IVHSs. If you said you would detail the extent of flooding in three areas, readers would be alerted to the end if you simply devote the last section to the third area. If your information is complex or extensive, however, you may end with a summary. The summary adds no new information but instead pulls together the report's major points.

IMRAD Report

If your information report describes empirical research such as a laboratory report or a clinical report, your audience probably expects a common structure for such reports, called IMRAD, which is an acronym for the report's components: introduction, materials and methods, results, and discussion. The box shows some important questions to ask when composing in an IMRAD structure.

QUESTIONS TO GUIDE THE ORGANIZATION OF AN IMRAD REPORT

Introduction
- Who asked me to look into this?
- Why did they ask *me* or our team?

- Why did they see this as a problem?
- Who else knows about this and has written about it (review of literature)?
- Has anything happened like this in the company or has been recorded in the literature before?

  o   If so, what was the outcome then?

- What limits did I have in solving this (budget, time, overlap with someone else's responsibilities)?
- What priority does this have in the company?
- What was I specifically asked to do?
- Are there any hidden agendas?

Materials and Methods
- Where did I look for information?
- What chief areas did I study?
- Whom did I talk with?
- What did I read?
- What surveys or observations did I make?
- What tools or technologies did I use?

Results
- What results did I obtain from my work? What did I find out?
- Are these results accurate? How do I know?
- Are these results valid? How do I know?

Discussion
- Do the results show any trends? Short term and long term?
- What do they add up to?
- How do they relate to other findings?
- Do the conclusions match my assignment?
- Have I overlooked anything I was asked to do?
- So what? In the reader's terms, and the organization's terms, what does all this matter? Future work?

Introduction

To draw attention and orient the reader to an IMRAD report, use the introduction to create, as one group of experts defines it (Swales, quoted in Harmon & Gross, 1996), a research space. Do this in three moves:

- Move 1: Define a research territory
- Move 2: Stake out a niche within that territory
- Move 3: Occupy or defend that niche.

Begin by setting the context—the static picture. Note prior research (that's the review of the "literature") to establish the scope and importance of the problem.

Then cite what disrupts that static condition—a gap, an inconsistency, a piece that doesn't fit the puzzle—and explain why that disruption is undesirable. Finally, note how or perhaps why your research resolves that problem and then lead the reader into the report to find out.

Materials and Methods

To assess the validity of your findings, your reader will expect a detailed discussion of how you conducted your work. The following paragraph begins such a discussion in a report on the design of a windmill:

> First, I obtained wind data from the weather station. The data were used to determine wind velocities and the distribution of hours of useful wind throughout the year. I then studied the efficiency of three types of wind wheels—multivane, turbine, and propeller—in extracting energy from the wind. In particular, I looked at the airfoil cross sections of a number of propeller blades. These data provided information for the design.

Results

You are likely to present your results in quantitative or visual form—often a series of tables or charts. You may also use diagrams and photographs. If appropriate, compare your results with expectations based on theory or published research. Note any limitations on your findings, including statistical reliability, any bias in the sample, or inconsistencies in data.

For more about visualizing data, see Chapter 9

TREATING DATA IN FORMAL REPORTS

In an analytical context, raw data (survey results, interview narratives, observation notes, etc.) are processed into categorical or thematic findings. They can be *descriptive* (e.g., participant ages, customer ratings of product, number of clicks on a link) or *interpretive* findings (e.g., common issues reported in customer reviews, frequently mentioned topics in an online user-support center). Through different analysis methods, such as statistical, textual, or qualitative treatments, data can be turned into meaningful insights. These insights help explain or predict a situation.

In reports, you rely on data-driven findings and insights to support your arguments or decisions. To achieve desirable outcomes, ensure you use the following practices in data treatment:

- Collect data from representative samples
- Collect enough data to produce generalizable findings
- Apply appropriate analysis methods
- Explain the meanings and significance of your findings
- Give enough discussions to help readers make an informed judgment.

Discussion

The discussion section shows what the results add up to, their implications and significance. You "occupy the niche"—that is, you show how your findings close the gap, resolve the inconsistency, add a piece of the puzzle, or solve the problem. You provide your conclusions about the evidence. Often, you trace a pattern of cause and effect, as in these segments from the discussion section in a report on a major coastal storm:

> The severity of the problem at Dewey Beach is in part due to the lack of a dune system in front of the buildings, the progressive loss of beach and sand due to beach migration inland, the lack of a significant natural source of sand in the surf zone and just offshore to naturally replenish the beach, such as is found at Rehoboth Beach, and the impact of a migrating coastline on structures not designed to be affected by waves and surf.
>
> (Ramsey et al., 1993, p. 22)

> As stated in the report of January 4, 1992 (DGS Open File Report No. 36), the worst-case scenario for a coastal storm impacting Delaware's coast that would produce severe and life-threatening conditions is as follows:
>
> 1. A slow-moving storm with tropical-to-hurricane-force winds
> 2. Landfall over the southern Delmarva Peninsula that places the Delaware coast in the storm's northeast quadrant
> 3. Continuation of the storm over several tidal cycles
> 4. Landfall during high tide or an astronomical high tide.
>
> (Ramsey et al., 1993, p. 23)

List your conclusions or present them in paragraphs. Listing makes the value of each conclusion seem equal. Paragraph form allows you more ease in discussing relative emphasis and priority.

After you have drafted your discussion, make sure it presents your information—and you—in the best light and meets your readers' needs. Use headings and other design devices to speed your reader through your report, and adapt your expression, as in other documents, to the readers' expectations and level of expertise.

Recommendation Reports

In Chapter 14, on "Business Plans," we introduced the feasibility plan as a document that helps decision-makers assess and determine if an idea, method, or procedure is viable for a project. Recommendation reports are based on the findings from a feasibility study and provide a specific course of action.

To persuade the reader to accept your recommendation, you need to create an effective argument. Let's assume, for example, you think demand management is the best transportation approach. If you anticipate that the reader will welcome, or at least not oppose, that recommendation, then use this organization:

- Introduction: brief overview of the problem and statement of the recommendation: demand management

- Description of the criteria for making your recommendation
- Brief review of each rejected alternative, along with reasons for rejecting it, in order from the least likely to the most likely
- Longer discussion of demand management (the most likely)
- Definition of terms, if needed
- Detailed analysis of why each method of managing demand will work
- Parking policies
- Employee commute plan
- Final statement of the recommendation: initiate demand management practices.

If you anticipate disagreement, consider withholding your final recommendation until you have detailed all the alternatives and convinced the reader to agree that only the last one you present—demand management—will work.

Introduction

The introduction to a decision-making report, then, draws attention either directly or indirectly, depending on your expectation about how the reader will respond. Here, for example, is a direct introduction to a recommendation report:

> John X. Fancher requested that Enviro Consultants Ltd. prepare a remodeling plan to conserve heat in his family's 50-year-old farmhouse. This report describes the plan, which is based on a thorough examination of the house. Recommended improvements should retain heat for longer periods, lower the fuel bills, and increase his family's comfort.
>
> The plan indicates recommended actions, but it does not discuss in detail what methods or materials to use. For these reasons, it does not include specific cost estimates.
>
> The report is divided into two parts: the causes of heat loss and recommended solutions to alleviate these problems. Briefly, the causes of heat loss are heat transmission through the building materials and air infiltration (passage of air through openings in the structure). Transmission of heat is simple to understand, so most of Part 1 centers on air infiltration.
>
> Part 2 details the recommended solutions in seven categories:
>
> Weather stripping   Vapor barrier
> Storm windows and doors   Windbreak
> Caulking   Entrance foyer
>   Insulation

Criteria

In Figure 16.1, the writers retrace their analysis process as backing for their recommendations. They state their criteria for the evaluation in the introduction and through the specific heuristics they include as the key to their web analysis method. In any decision-making report, you need to include such criteria. Do so in the introduction, if the list is not long, or in a separate section if their extent warrants.

**WEBSITE HEURISTIC EVALUATION
AND RECOMMENDATIONS REPORT**

Prepared for

Dr. Sukant Misra
Vice Provost for International Affairs
Office of International Affairs
Texas Tech University

Prepared by

Megan Dawson
Web Usability Specialist
Digital Solutions, Inc.

November 9, 2021

Include report title and recipient information on the cover page

Introduction

Start with a summative introduction

Website usability can influence user experience and overall satisfaction toward services offered by an organization. At the request of Dr. Sukant Misra from the Texas Tech University's (TTU) Office of International Affairs (OIA), I evaluated the current homepage of the office and provide redesign recommendations. This report presents the results from my heuristic evaluation (see Methods below) and summarizes areas for improvement on the OIA website.

Methods

Provide details about methods if your report follows a particular methodology

This document presents the analysis of the TTU OIA website at Texas Tech University using usability expert Jakob Nielsen's "Ten Usability Heuristics for User Interface Design" (1994). I tested the desktop view of the website on the Mozilla Firefox browser and a HP Pavilion laptop. Dr. Misra had expressed interest in creating a positive experience for current international students and potentially new applicants with the OIA website. By applying Nielsen's 10 heuristics, I analyzed different aspects of the OIA home page, made suggestions for improvement, and added a corresponding screenshot for each heuristic. At the end of the report, I included a summary table (Table 1), which provides a rating of each heuristic according to Nielsen's Severity Rating scale, and an overview of its strengths and weaknesses.

Include page numbers throughout report

1

16.1
Recommendation Report on the Evaluation of a Website
Source: Courtesy of Meghalee Das

1. Visibility of system status

Use graphical elements to distinguish key elements in your report

Heuristic description: The system should always keep users informed about what is going on, through appropriate feedback within reasonable time.

Figure 1. The OIA home page with icons and menu informing users and giving them feedback.

The OIA home page gives information about different functions as seen in Figure 1 above. There are notifications for important news such as COVID-19 updates, with the option to click for more information (A). When the cursor is moved over social media icons on the left (B), they extend slightly on the right and display the name of the social media site. Similarly, icons on the lower menu bar (C) become dark, and the top menu bar (D) become deep red when the cursor hovers over them. But there is inconsistency in visibility of feedback. For e.g., the OIA top menu bar icons (C) don't say the name of the page. This can be confusing to international students, who may not recognize the icons.

2. Match between system and the real world

Heuristic description: The system should speak the users' language, with words, phrases and concepts familiar to the user, rather than system-oriented terms. Follow real-world conventions, making information appear in a natural and logical order.

Whenever appropriate, use figures to demonstrate your argument

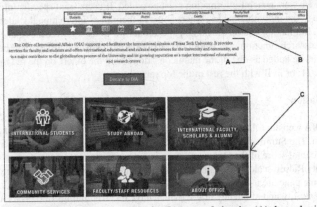

Figure 2. The sentences need to be shorter and simpler (A), but a logical order between B and C is followed.

2

16.1
(Continued)

The OIA home page uses fairly clear language that should be familiar to most international students. However, it should be tested with a Simplified English checking software, which can help non-native English speakers to understand the content, and make translation easier if they choose to do so. For e.g., in Figure 2, the second sentence should be broken up into two sentences at least, as according to Simplified English guidelines, a sentence length should be around 20-25 words long (A). The information appears in a logical order in most areas of the home page. For e.g., the order of the links on the menu bar on top (B) matches the six icon boxes in the lower half of the page (C). This logical order will make it easier for the user to find the information. However, the Scholarships page is missing, so that should also be included somewhere below to keep the same order.

10. Help and documentation

> Heuristic description: Even though it is better if the system can be used without documentation, it may be necessary to provide help and documentation. Any such information should be easy to search, focused on the user's task, list concrete steps to be carried out, and not be too large.

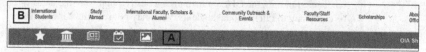

Figure 11. A general FAQs or help section/icon must be added in one of the menu bars.

The OIA website is complex because it has so many functions and services. But there is no help icon on page that users can click to get basic information. An FAQs icon or menu item which answers general information must be added. If users want specific information, they can go to the individual pages, where some of them like "international student life" have FAQs. But having one on the home page about how to use the website, make appointments, use the search function, and whom to contact for each key department will be helpful. That icon can be added as one of the mainicons along with the others on the red menu bar (A) in Figure 11, or in B with the rest of the top menu items.

Conclusion

Close with a summary of findings and recommended actions

The OIA website follows most of the Nielsen's Usability Heuristics for User Interface Design. The Summary Table in the next page (Table 1) contains the ratings and summary, and the total score of the OIA home page is 2. On a scale of 0 to 4 according to Nielsen's Severity Rating scale, with 0 meaning no problems, a score of 2 is good. The strengths of the home page are user control and freedom, error prevention and confirmation messages, and flexibility and efficiency of use, which have received ratings of 0 or 1. Guidelines on visibility, match between system and real world, and standards are also fairly followed, with

12

16.1
(Continued)

some recommendations like the use of Simplified English and more consistent layout. In terms of aesthetics also, a more minimal style with fewer graphics will be more effective. But the most important area of concern is that there is no Help or FAQs section or icon on the home page. This might be due to the complexity of the services and so they might have been added to individual pages. But having a general FAQs and Help section prominently displayed on the home page will be a good idea, especially since some international users may not be fa-miliar with where to start looking for information. Nevertheless, the OIA home page is designed effectively, and by executing some of the recommendations, it can develop into a more user-friendly website.

A table is suitable for providing an overview of results

Table 1. An overview of the strengths, areas of improvement, and rating for each heuristic applied to the OIA website using Nielsen's Severity Rating scale. The ratings are in a scale of 0-4, with 0 meaning no problems, and 4 meaning severe problems.

Heuristics	Strengths	Areas of Improvement	Rating*
1. Visibility	Some feedback through immediate color change or page name appearing	Consistency needed in feedback display in all links and icons	2
2. System-Real World Match	Logical order followed between content and menu throughout the page	Simplified English language and shorter sentences should be used	2
3. User control and freedom	Emergency exit, and new window opens when links and icons are clicked		0
4. Consistency and standards	Consistency in font and TTU website standards	Consistency ideas on layout and style of the page can be explored	2
5. Error prevention	Error and confirmation messages for services like Make an Appointment		0
6. Recognition	Breadcrumbs, search, index are useful	More study on other pages needed	1
7. Flexibility and efficiency of use	Multiple links of most used actions and options; some keyboard shortcuts	Double click appointment button; shortcut register on events page	3
8. Aesthetic and minimalist design	Colors are well balanced; easy to navigate with visible menus and icons	Reduce number of graphics and use cleaner layout	3
9. Help recognize, recover from errors	Some pages and forms highlight missing information to help recognize errors	Add suggestions to help users recognize and recover from search errors	3
10. Help and documentation	Some individual pages have FAQs	General FAQs or help section/icon must be added in menu	4
		Average	2

13

16.1
(Continued)

Conclusions and Recommendations

Your examination of the alternatives leads to conclusions on which you base your recommendations. A reader can often predict your recommendations based on your conclusions, but they are not the same. Conclusions look to the past (e.g., what was done in a study); recommendations look to the future (what to do next). Recommendations are the lifeblood of decision-making reports. Here is one segment from the recommendations section of a report on "Growth in a Global Environment":

> The Global Climate Coalition recommends a policy including the following options:
>
> - Accelerate the pace of the research into basic climate science and impact assessment
> - Identify and pursue measures that will reduce the threat of climate change, yet also make sense in their own right
> - Establish sustained research and development programs that improve the ability to economically produce and utilize energy with less potential for the accumulation of greenhouses gases
> - Expand efforts to understand and communicate the economic, social, and political consequences of both climate change and the proposed policy responses.

After you have drafted your discussion, make sure it presents your information— and you—in the best light and meets your readers' needs. Use headings and other design strategies described in Chapters 9 and 10 to guide your readers through your report, and adapt your expression, as in other documents, to the readers' expectations and level of expertise.

IDEATING AND PROTOTYPING SOLUTIONS

Design thinking can help you determine the cause(s) of a problem you're facing and devise viable solutions. The ideation mode in design thinking encourages divergent thinking and radical imagination. When creating recommendation reports, consider using the CIR approach:

- *C—conventional*: Propose solutions that abide by the existing expectations of the organization; try not to surprise or make waves by recommending "safe" ideas.
- *I—improved/innovative*: Suggest solutions that take conventional ideas one step further. Innovative ideas are those that follow organizational values and traditions but take on a modern twist. They often spark productive questions and excitement.
- *R–radical*: Recommend out-of-this-world solutions juxtaposed with practical solutions. The goal of doing so is to inspire stakeholders to consider pursuing something they never thought to be possible or doable. Although radical ideas rarely get taken up, they can lead to inventive solutions that defy typical assumptions.

THE REPORT PACKAGE

To fulfill their obligation to the reader, some reports need to incorporate additional elements beyond the discussion itself. The discussion becomes part of a larger package that includes a letter presenting the report to the reader, a title page, an abstract or executive summary, and references. In this section, you will learn about preparing the other elements you find in Figure 16.1, along with additional elements: a foreword or preface, glossary, and appendix. These elements accompany formal reports, often those written from one organization to another. Internal reports usually circulate in memo form and rarely require such extensive treatment.

Letter or Memo of Transmittal

The letter or memo of transmittal is a permanent record of your delivery of the report to the reader (Figure 16.2). Sometimes, a standard form is appropriate, with individualizing information filling in the blank spaces. But you can also use the occasion to converse with the reader. In your letter (or memo, if the report is circulating inside the organization), you announce the subject or title of the report, the authorization, and the date of authorization. In addition, you may

- Briefly discuss the content of the report
- Note for this particular reader the most pertinent sections (perhaps those the reader assisted with)
- Offer to answer questions about the project
- Remind the reader about any changes in scope of approach since the project was approved
- Acknowledge the assistance of people who helped you prepare the report.

Attach the letter to the report's cover or include it as the first page inside the cover. The length of the letter depends on the extent of explanation or courtesy needed. The tone reflects your relationship with the reader, from informal to formal. It's usually best to take a formal approach if you are unsure.

Title Page

The title page presents the title, of course, along with the names of the authors and the audience, the date, and any other necessary information about the report's origin and use. Your reader notices the title first, although you may consider it almost an afterthought. Take time to create a good title. Be brief, clear, and comprehensive. Use terms your intended readers would use to search for your report in an information retrieval system and place those terms up front. Place terms such as *final report* in the background. Here are some effective titles:

- *Bayside Tract Management Plan* (the proper name of the location precedes the more generic term, management plan)
- *Supporting Research and Development in the NHS: A Report to the Minister for Health*
- *The Health of the UK's Elderly People.*

16.2
Letter of Transmittal
for Website
Analysis and
Recommendation
Report

DIGITAL SOLUTIONS

123 Fall Avenue
Madison, WI 53716

Dr. Sukant Misra
Vice Provost for International Affairs
Office of International Affairs
Texas Tech University
Lubbock, TX 79409

Dear Dr. Misra:

Per your request, I have conducted a heuristic evaluation of the Office of International Affairs (OIA) homepage. In the attached file, you should find my evaluation report with redesign recommendations.

Overall, I found the OIA homepage to follow most of the Nielsen's Usability Heuristics for interface design. I noted the strengths of the current homepage design and two major redesign considerations—help documentation and simplified language for better accessibility.

I welcome your feedback on my report and look forward to meeting with you to discuss next steps.

Yours sincerely,

Megan Dawson

Megan Dawson
Web Usability Specialist
Digital Solutions, Inc.

Avoid being cute or punning (save such titles for popular articles):

- Report: *Recommended Practices to Prevent Traffic Accidents on Freeways*
  Article: *Mass Murder on the Freeways*
- Report: *Recommended Removal of Interior Wall*
  Article: *Wall of Conflict.*

A title that depends on jargon or terminology explained in the report is not sufficiently informative. *An Explanation of Pips and Poops in the HRT* may at first glance be amusing, even intriguing, but it is meaningful to only a very few specialists. More appropriate is *An Explanation of Sudden Power Surges and Drops in a Homogeneous Reactor Test.*

Preface or Foreword

A preface or foreword may accompany an extensive or a controversial report, especially one that is bound and published. Conventionally, a preface is written by the report's author, whereas a foreword reflects the views of another authority who comments on (and usually praises) the report. It may call attention to some unusual aspect or indicate particular problems in dealing with the investigation, and it some-times acknowledges help. The foreword discusses the qualifications of the author and the effectiveness of the report.

Glossary

A glossary acquaints the reader with the meaning of technical terminology, abbre-viations, and symbols (if any) used in the report. Some authorities discourage the use of glossaries because they require the reader to flip back and forth though the report. If you decide to include a glossary in your report, you can place it right after the table of contents so that all readers see it, or you can put the glossary in an appendix so it is available to those who are not familiar with the technical terms used but not in the specialist's way. Glossaries help you accommodate a wide range of readers with one document, but they should not substitute for a careful approach to using and defining terms in the text. Glossaries are particularly helpful in documents that will be translated.

References and Appendix

For more about incorporating sources, see Chapter 7

Two other elements you may include in a final report are a list of references (or bibliography) and an appendix. If you cite the literature in your report, properly credit the sources from which you borrowed information, as in the reference section at the end of each chapter in this book. An appendix furnishes supplementary material not essential to the development of the report that would, if included in the discus-sion, interrupt continuity. A properly labeled appendix is an appropriate place for

- Complete tabulations of data
- Sample calculations
- Detailed quotations
- Copies of letters that add to the validity of the evidence.

DELIVERING ONLINE REPORTS

Many companies and agencies use digital information management systems to create and deliver their reports internally. Externally, they rely on web platforms to share reports to customers and clients. Online reports reach readers more rap-idly than traditional print documents and, at the same time, reduce the costs of production, storage, and consumption of natural resources associated with paper and printers.

Writers may face new challenges in preparing online reports. A simple and imperfect solution is to "dump" the paper document into a web application, ignoring most, if not all, of the differences that readers may encounter between paper-text and text on a screen. But such ignorance is not bliss for the reader. Writers of paper documents highlight information and ease readability through page design and textual elements such as tables of contents, indexes, and abstracts. It's easy for readers to skim a paper report to get its gist and find information they need.

Writers of online reports have to compensate for the size of screen (desktop versus mobile screens) and difficulties in skimming using different designs that help readers navigate through contents. Online modules (partitioning a report into separate sections on a webpage) can offer some compensating features, including search and retrieval options, links, and help documentation. These features let readers find information not just from one report, but from a database that includes a series of reports to create the one compendium of information needed to support a particular decision. Properly used, online delivery makes it easier to customize a report as you segment information in units that different readers can use in different ways.

Formal reports supplement your project by communicating results (information report) and recommendations (decision-making report). Depending on your situation, structure formal reports using specific organization for explanatory, IMRAD, and recommendation reports. When delivering a report, supplement the package with a letter of transmittal, front matters (title page, preface or foreword), and back matters (glossary, references, or appendix). Consider paper and screen differences if you choose to deliver your report online.

CHECKLIST: FORMAL REPORTS

1. Determine your purpose for writing the report
   - Write an information report to help your reader understand a concept, item, or situation
   - Write a decision-making report to help your reader solve a problem, make a decision, or act in the right way

2. Select information that will support your purpose and your reader's needs
   - Respond to any specific request from the reader
   - Think of questions that will require answers

3. Structure your report to conform to the genre appropriate to the context
   - In the abstract or executive summary, highlight the content or findings that require most attention
   - In the introduction, orient the reader to the focus of your report
   - To describe a concept, situation, event, system, or object, use an explanatory report
   - To describe empirical research, use an IMRAD report
   - To lead the reader to a decision, use a recommendation report

4. Test your report's approach, structure, and appropriateness
   - Conform to the accepted conventions of appropriate genres
   - Meet the audience's needs

5. Include supplementary elements as needed

EXERCISES

1. Construct a purpose statement for a decision-making report and a thesis statement on the same subject for an information report.

2. Identify an article from a scientific or technical journal in your field. Write an executive summary (business or government publication) or abstract (academic or research publication) for the article. If the article already has an abstract or summary, assess its effectiveness by judging whether it highlighted the article's purpose, methodology, results, recommendations (if any), and conclusions.

3. Using the web evaluation report example in Figure 16.1 as a model, perform a design analysis on a website of your choosing. Use the basic principles of web design (Nielsen, 2020) below as your guidelines. Then, create a recommendation report addressed to the webmaster of your selected website.

   - *Visibility of system status*: The design should always keep users informed about what is going on, through appropriate feedback within a reasonable amount of time.
   - *Match between system and the real world*: The design should speak the users' language.
   - *User control and freedom*: Users often perform actions by mistake. The design should support undo and redo.
   - *Error prevention*: Eliminate error-prone conditions or check for them and present users with a confirmation option before they commit to the action.
   - *Recognition rather than recall*: Minimize the user's memory load by making elements, actions, and options visible.
   - *Flexibility and efficiency of use*: Include shortcuts—hidden from novice users—that may speed up the interaction for the expert user.
   - *Aesthetic and minimalist design*: Interfaces should not contain information that is irrelevant or rarely needed.
   - *Help users recognize, diagnose, and recover from errors*: Error messages should be expressed in plain language (no error codes), precisely indicate the problem, and constructively suggest a solution.
   - *Help and documentation*: It's best if the system doesn't need any additional explanation. However, it may be necessary to provide documentation to help users understand how to complete their tasks.

FOR COLLABORATION

In a team of two or three, write a formal report, either informational or decision-making, on a topic of your choice that is approved by your instructor. Review Chapter 5 concerning categories of problems and then pick a topic that is interesting enough to sustain your team through several weeks of research and writing. Ask questions. Establish the boundaries of your report by setting down the subject, problem, purpose, procedure, scope, and audience. You might include such information in a formal proposal (see Chapter 13). Here is an example:

> *Question*: Will maintaining two rights-of-way connecting the same cities be economically feasible?
>
> *Subject*: Conrail's request filed with the Interstate Commerce Commission pertaining to the abandonment of trackage in eastern Ohio.
>
> *Problem*: Conrail operates two parallel branches from Youngstown, Ohio, to Ashtabula, Ohio; both lines are in need of extensive repair owing to lack of maintenance and funding.
>
> *Purpose of investigation*: The purpose is to determine whether maintaining two rights-of-way connecting the same cities is economically feasible, and whether the operation of a single track will free money for maintenance that is now tied up in the operation of a second line.
>
> *Procedure and content*: The report will interpret the results of a survey concerning total daily train traffic on each of the two lines. These results will be used in economic analysis emphasizing that the same total revenue can be realized by the use of one track daily instead of two. At the same time, daily maintenance costs can be cut as much as 48 percent. The success of a "smart" train-scheduling method will be described to further justify the request. The method maximizes safety in the use of a one-track right-of-way.
>
> *Audience*: The Interstate Commerce Commission (primary audience), Conrail officials (secondary audience).

REFERENCES

Eccleston, C.H. (2014). *The EIS book: Managing and preparing environmental impact statements* (Chapter 6). New York, NY: CRC Press.

Harmon, J.E., & Gross, A.G. (1996). The scientific style manual: A reliable guide to practice? *Technical Communication, 43*(1), 61–72.

Nielsen, J. (2020). 10 usability heuristics for user interface design. Nielsen Norman Group. Retrieved from https://bit.ly/3elcG0I

Ramsey, K.W., Talley, J.H., & Wells, D.V. (1993, February). Delaware geological survey (DGS). *Summary report, the coastal storm of December 10–14, 1992, Delaware and Maryland*. Open File Report no. 37.

17 Instructions

Technical communication involves conveying technical information to non-expert audiences so they can achieve desired goals. Therefore, instructions are a staple genre in technical communication. As a technical professional, you'll spend a good deal of time reading and writing instructions for making things work. As a static or moving visual, as written text, or as both visual and text, such instructions appear

- On a device
- On the box containing a device
- On a tag
- In a memo, brochure, or manual
- Online, in a help window or user forum.

Instructions pertain to products (such as toaster ovens), systematic approaches to behavior (such as student codes of conduct), and methods (such as conducting a test in a laboratory). They may be called by a variety of names—for example, procedures, directions, and directives. The document that contains them is often called a manual. All instructions divide an action to be performed into steps the reader can manage. In this chapter, you will learn guidelines for instructing your user.

PURPOSE

You have two related purposes in creating instructions. First, your instructions enable the reader to use the product, behave appropriately, or follow the method to achieve the desired result. That's the obvious purpose. In addition, instructions persuade the reader to buy or behave or follow through an explanation or description. This second purpose is one technical communicators forget at their peril.

Enabling Use

To achieve the first purpose, think of the reader of the instructions as a product user. Ask these questions: What problem is the product or system designed to solve for them? How will they measure success? What preventive measures should

DOI: 10.4324/9781003093763-21

they take to avoid damage or harm? The following statement from an Apple Watch instructional page teaches the user about measuring heart rate:

> The optical heart sensor in Apple Watch uses what is known as photoplethysmography. This technology, while difficult to pronounce, is based on a very simple fact: Blood is red because it reflects red light and absorbs green light. Apple Watch uses green LED lights paired with light-sensitive photodiodes to detect the amount of blood flowing through your wrist at any given moment. When your heart beats, the blood flow in your wrist—and the green light absorption—is greater. Between beats, it's less. By flashing its LED lights hundreds of times per second, Apple Watch can calculate the number of times the heart beats each minute—your heart rate. The optical heart sensor supports a range of 30–210 beats per minute. In addition, the optical heart sensor is designed to compensate for low signal levels by increasing both LED brightness and sampling rate.
>
> (Apple, n.d., n.p.)

Explaining to Persuade

To achieve the second purpose, consider how your document is designed to persuade users to take actions. Good instructions provide the rationale for using a product or system in a certain way so the users can achieve optimal benefits or success. The explanations in the instructions should convince users to *follow* the prescribed steps or methods for using the product. As such, the explanations should appeal to the following elements of persuasion:

For more about persuading versus explaining, see Chapter 8

- *Credibility of the provider*: establishing the authority to give instructions; presenting expert insights and ethical treatment of information
- *Interests of the user*: making users care about the instructions by showing how they would benefit from them
- *Precision of the instructions*: ensuring factual accuracy and logical descriptions; using strong evidence to support instructions
- *Timeliness and appropriateness*: demonstrating awareness of the use scenarios and proper (socially acceptable) approaches.

INSTRUCTIONAL SITUATIONS

Instructions are needed in various situations. The three common situations are descriptive processes, prescriptive processes, and blended descriptive and prescriptive processes (Fleming, 2020).

Descriptive processes answer the question, "How can this be done?" These instructions describe how exactly a process happens. For example, in Figure 17.1, the internship guide for students in a wind energy program describes the steps to setting up an internship and the necessary materials.

Prescriptive processes answer the question, "How should this be done?" These instructions give users the knowledge to perform certain tasks so they can

17.1
Sample Instructions on a Website Guidelines for establishing an internship by undergraduate students in the Wind Energy Program at Texas Tech University.

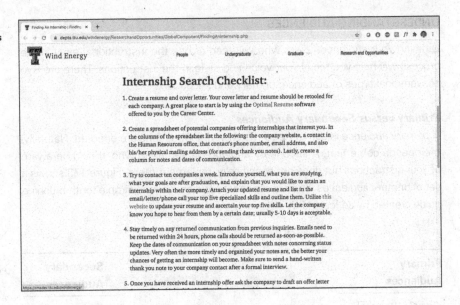

17.2
Blackboard Troubleshooting Tips

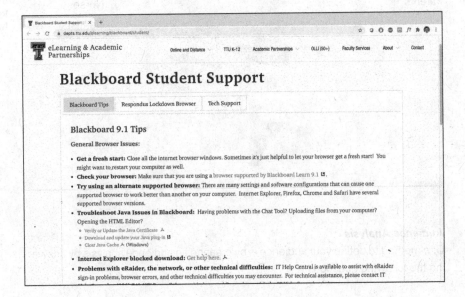

do those themselves. Prescriptive instructions are explanatory in nature. Figure 17.2 shows an example of "tips" to using the learning management system Blackboard at Texas Tech.

Blended descriptive and prescriptive processes are most common for technical instructions as they combine the "what" and "how" of doing something. Within the same document (but in different parts), the blended instructions give information about the product (e.g., what it's made of) and teach users how to use them (e.g., how to assemble a trampoline).

UNDERSTANDING AUDIENCES

Defining your audience (users who will read or use the instructions) early in the process will help you focus your voice and style in the instructions. There are two conventional types of audience: primary and secondary.

For more about diverse audiences, see Chapter 2

Primary versus Secondary Audiences

A *primary audience* is someone to whom your instructions are directed. Naturally, their needs come first. A *secondary audience* is someone who should be aware of your instructions but does not specifically implement them. Figure 17.3 shows a set of primary audiences vis-à-vis secondary audiences for a simple tooth-brushing guide created by an Icelander.

Primary Audiences		**Secondary Audiences**
Kids (4–12 years)	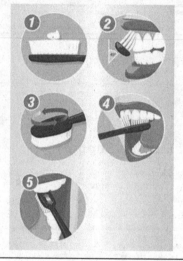	School teachers
Parents		Nurses
		Dentists

17.3 Primary vs. Secondary Audiences A tooth-brushing guide. Retrieved from https://bit.ly/3gqlzlU

Audience Analysis

One method to define your audience is by creating an audience analysis by answering the following questions:

1. *How homogeneous is the audience?* Can you assume all your readers share major characteristics—motivation, expectations, prior knowledge, and the like? Are you addressing a small group of experts? A large group? How diverse is the group?
2. *How widespread are your users geographically?* Will the manual supplement personal instruction or must it stand on its own for readers who will need to use it at a distance from your company's support system?
3. *What constraints pertain to your document?* For example, what legal and commercial standards and codes apply in the markets where you'd like to sell?

4. *What is the user's technical level?* Will the document need to teach technical concepts or principles before it can describe the tasks at hand?

5. *What is the user's preferred language?* If the answer is a language other than English—and you are writing in English—you'll need to build in special features. For example, readers may be more comfortable with technical terms in English than with colloquial English.

6. *Will the manual be translated?* Prepare the document for possible translation. Such preparation helps companies release products simultaneously throughout the world, even if the manual has to be localized—that is, translated for a particular country or language group. Translating is expensive, and that expense can usually be justified only if the market in a particular country is large. For most technical products, instructions in an international language such as English are adequate, so long as that English is simple and precise.

7. *How much detail do users need?* Are your readers already familiar with similar devices or systems so you only need to update them on differences? What expectations do they have about the extent of discussion in a manual?

8. *What technology do users share with you?* Can they download your manual if you publish it on the web? Can they open a file sent as an email attachment? Is a print manual the only form they can access? Is it their preferred form?

You may use your audience analysis to adjust your approach to overcome any resistance you anticipate from the reader. Some people readily adopt new ideas or new technologies; others resist. Because of their high motivation, adopters make fewer demands as readers than resisters. For resisters, be imaginative in the use of visuals, design elements, and interest-getting devices. The following examples summarize briefly the answers to some of these questions for a particular set of instructions. The first is for users of home medical devices, and the second is for users of a software help system. Table 17.1 shows a more extensive analysis in a scientific setting.

- Users of home medical devices range widely from the sophisticated to the less literate. Many are elderly; for many, English is not their primary language; some are on medication that impedes attention span and memory; some have poor vision; some have poor hand–eye coordination (Backinger & Kingsley, 1993).

- Customers for the help system probably fall into at least three categories: new users, experienced users who need guidelines for a seldom-used program, and advanced users who need reference information and programming details (McLaren, 1993). New users need basic information to get them started, perhaps with a comparison to a program or activity they already know. Any user may need a reminder about something forgotten, methods for correcting an action that led to trouble, orientation when they are lost, and a quick way out.

Table 17.1
Extensive Audience Analysis

What is our objective?	• Show potential users how the system can meet their needs and encourage them to try it out
Who are the main target readers?	• Scientists who could use the system to gather hydrology data remotely
Who else might read the guide?	• The scientists' assistants • Administrators who want to know how much it will cost • Data processing security people who want to check that any confidential data will not leak out from their sites
What are their main questions?	• Will the system offer easy access to data? • What are its unique capabilities? • What will guarantee the continuity of the system over the next ten years? • Who will help if they have operational problems? • Which members of their peer group are already using the system, and what do they think of it?
What else may readers have seen about the Toulouse System?	• The Toulouse newsletter • Scientific papers mentioning the Toulouse System • Advertisements on professional journals' websites
Likely reading pattern?	• Quickly skim through introductory segment on theory of operation and history • Read list of hydrology capabilities more closely • Look up specific issues in table of contents or index
Lessons to draw from their reading pattern?	• Make each capability a separate, self-contained segment • Use smaller than A4 page, brochure format • Use plenty of tables, lists, and illustrations

Note: A technical writer in France created this analysis of the audience for the hydrology version of a commercial geographic information system, the Toulouse System

Source: Courtesy of Nigel Greenwood

STRUCTURING INSTRUCTIONS

Clear instructions should appear to be simple and easy to follow. Keep the Murphy Law for technical instructions in mind: if something can be misunderstood, it will be misunderstood. It's particularly easy to be misunderstood if your readers vary in skill level, differ from you in corporate or national culture, or represent a variety of different cultures. Design your instructions to fit your analysis of the readers. In doing so, include some or all of the segments described in this section.

Introduction

An introductory unit to such instructions as package inserts, manuals, or sets of guidelines welcomes the reader and overcomes any reluctance to read. A brief introduction to instructions for assembling a device may, for example, list pieces included in the box, overview the major steps ("You can assemble the bed frame in three easy steps"), and establish a time frame ("This assembly should take one hour"). More detailed instructions for using a device or system may include a longer introduction, as in the box on "A Welcoming Introduction in a User Manual." In such a segment, often headed "Read Me First" in U.S. publications, you achieve some of the following purposes:

- Reiterate the purpose and importance of the device or system
- Explain how the manual is organized
- Establish the level of special knowledge or skill required
- Identify the intended user of the device or system
- Provide a sample scenario of the device's applications
- Instruct the user on how to read the manual.

A leading technical professional in China notes that Chinese readers expect the introduction to help build trusting relationships between producers and customers in a culture where such personal relationships are an essential part of doing business. The introduction should "establish the image of the enterprise, emphasizing the company history, business activity, range of products, and their application" (Zhu, 1995, p. 11).

A WELCOMING INTRODUCTION IN A USER MANUAL

Congratulations

on your new bicycle! Proper assembly and operation of your bicycle is important for your safety and enjoyment. Our customer service department is dedicated to your satisfaction with Pacific Cycle and its products. If you have questions or need advice regarding assembly, parts, performance, or returns, please contact the experts at Pacific Cycle. **Enjoy the ride!**

About This Manual

It is important for you to understand your new bicycle. By reading this manual before you go out on your first ride, you'll know how to get better performance, comfort, and enjoyment from your new bicycle. It is also important that your first ride on your new bicycle is taken in a controlled environment, away from cars, obstacles and other cyclists.

Source: Mongoose Mountain Bicycle owner's manual

Theory of Operation

Your audience may also expect a unit in the instructions providing the theoretical underpinnings for the process. German and Dutch readers, for example, like to see an extensive discussion on the theory of operation at the beginning of a manual. They tend to read manuals carefully before performing tasks and need theory as context for those tasks.

Storytelling

Storytelling is a method for understanding user needs and designing communication solutions that address specific requirements. As well, stories can be used in technical instructions to enliven the document with personality and voice. Consider the case of a food recipe. Many recipes—as instructions—include narratives about the food under preparation to engage their readers. Manuals and technical guidelines, too, may use storytelling as a way to create coherence and better user experience (De Silva & Henderson, 2007).

Safety

You should address safety concerns in a special unit near the beginning of a manual. Government regulations in Japan, for example, require manuals to begin with extensive warnings and cautions aimed at keeping operators safe while performing the process (Burnett, 1997). In the United States, too, manufacturers devote attention to identifying hazards, in part out of fear that they may be liable for damages. Note such safety issues at the beginning of the manual and at pertinent points throughout. Include a generally recognized icon that signals the problem, briefly explain the problem and the consequences if your instruction is not followed, and instruct the user in avoiding the hazard. The two major safety notations are warnings and cautions.

Warnings

Write a warning to draw attention to something that threatens the user's personal safety, including serious injury or even death. You may also use the term danger.

Instructions for some medical devices have to warn the user not to operate the device if certain symptoms or problems occur. They might also warn about backup procedures necessary if life support systems fail. Most warnings are set off from the text in boxes or accompanied by icons that grab the reader's

17.4
General Warning
Label

**17.5
Standard Caution
Icon**

eye. Standards for the icons, however, vary. The International Organization for Standardization (ISO) dictates the use of a triangle with an exclamation point for a chemical warning, and a triangle with a lightning bolt for an electrical one. Some documents use a red stop sign with a raised palm. The color disappears, however, if the manual is printed in black and white, so relying on color as a warning may be ineffective.

Cautions

Use a caution to alert readers to potential malfunctions or failures of the system itself and any resulting damage, including, for example, a loss of data if a computer system goes down. A safety precaution aims to guard both the equipment and the user against any misuse:

> Protect the printer against extreme cold or hot weather as well as dampness and wet conditions.

Cautions are sometimes indicated by the color yellow and by an exclamation point in a triangle (according to the IOS, ISO 7010-W001).

Conditions Affecting Use

If they are significant, explain the conditions required for operating a device or conducting a procedure. For example:

- The need for a source of water and electricity
- The need to unplug the device before testing or cleaning.

Similarly, clearly state any conditions in which the procedure should not be conducted or the device not operated:

> A hairdryer should not be used near water—near a bathtub, sink, wet basement, or swimming pool.

Tell the reader how to vary the procedure if the device operates differently:

- At various altitudes or temperatures
- In different geographic regions
- In transit between locations.

Prepare the customer to anticipate and correct any problems.

Setup

If appropriate, clarify whether readers should set up the device on their own or seek professional help for that step. If the readers will perform the setup, note any special requirements (Backinger & Kingsley, 1993, p. 9):

- Parts list
- List of materials and tools needed for setup
- Unpacking instructions
- Directions for locating the device in the home, such as on a tabletop or on the floor (also state whether the device should remain in one place after setup)
- Any warnings or safety instructions specifically related to setup, placed right before the corresponding task or instruction
- Description of what happens if setup is wrong
- Setup instructions in steps numbered in logical and chronological order
- Any special preparation before first use of the device, such as cleaning or disinfection
- Space to write in user-specific instructions
- Who to call if there is a problem.

In addition, explain, if necessary, any checks readers should run to ensure that the device is working and is properly calibrated.

Step-by-Step Description

At core, the instructions describe a series of steps readers must take to achieve the desired goal. Sometimes there is only one way to do something. But often, you can reach the same goal by different routes. Find the route that best matches your analysis of the readers so you reduce any initial resistance and uncertainty. Look for trouble spots as well as shortcuts. Establish the outcome of each stage.

One method for developing the sequence is to show an expert performing the task, usually via video, as in Figure 17.6. Use close-up and cut frames to identify discrete activities. In addition, interview experts who use the product—for example, the designers who developed it or major customers. In an interview, however, an expert may not think of some action that's now second nature. For example, an expert may think the first step is simple: Turn the machine on. But look for the possibility that more than one action is required, as in the following description of turning on a machine:

To turn the machine on,

1. Plug the power cord into an AC outlet.
2. Facing the front of the machine, find the black power switch on the right side.
3. Turn the power switch to the "ON" position.

To avoid the bias of an expert, you can work with the product or perform the method yourself, assuming you are not an expert. See what it takes to achieve the right outcome. You might also ask some novices to help you determine what steps are necessary. Suggest guidelines and then watch the novices perform the activity. Note, for example, if any materials or equipment are needed beyond what you originally planned. In addition, observe the following:

- How difficult the reader finds each step
- Whether any steps are significantly less or more difficult than others
- Where novices seem to go wrong in using the product or understanding the procedure
- How you can help the reader avoid those wrong turns.

Consider creating video instructions to demonstrate complex tasks, as in Figure 17.6. The multimodality of video instructions (i.e., visual + narration + animation + text) can afford better information processing. Users can watch and rewatch (or pause) the video to find the exact step in an instruction, which can help them perform tasks with greater accuracy (Poe Alexander, 2013).

Cleaning, Maintenance, Updates

In instructions for using devices, you may need to tell users how and how often to clean the equipment. Provide pullout or laminated maintenance charts that can be hung on the wall for easy reference. In addition, provide guidelines for storage (for example, "must be stored in a cool place" or "keep out of sunlight").

For digital products, your instructions should include ways the user can manage updates to the software or applications. Figure 17.7 shows an Apple macOS upgrade guide with specific hardware and feature requirements.

17.6
YouTube Video
Tutorial
Assembling quill
stem handlebars
on a Mongoose
bike. Retrieved
from https://bit.
ly/3n4ajTM

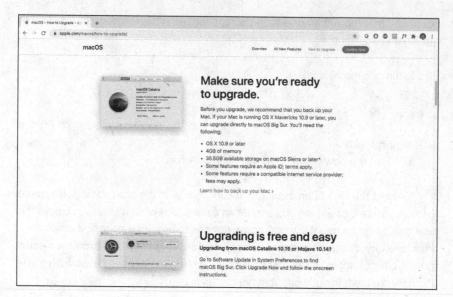

17.7
Maintenance or
Update Guides
Upgrading to
Apple macOS Big
Sur. Retrieved
from https://apple.
co/2QJwqD6

Troubleshooting

In your unit describing the steps in the process, you show what happens when everything goes right. In the troubleshooting section, you show what to do when the process goes wrong. Trial runs with potential users help you anticipate problems and prepare solutions. In addition, design your troubleshooting section for rapid reference. Provide a table or flow chart that includes signs of trouble and actions to counteract the difficulty:

- List error messages, if the device provides these, and then show how to correct the error. In any error messages, differentiate between what the system or product did and what the user did.
- Group related problems and use typography to highlight especially serious or common ones.
- Avoid lengthy technical explanations. As one veteran sailor noted in another context, "When the boat is taking on water, don't talk about navigation. Bail."
- Clarify what troubleshooting steps the user can complete, what problems the user should not attempt to remedy, and when and how to call technical support.

Other Items

In addition to these major units, your instructions may also include other items. For example, you may create a quick reference card for seasoned users. The card provides brief reminders of key steps or functions, along with cautions and warnings. Use a single page or a narrow column that can fit on the device itself. In addition, list prominently (preferably on the device itself) a tech support phone number, web link, or email and product support center addresses for customer service.

You may also include notes that qualify information in the text, provide suggestions for further information, add technical fine points, or offer hints to make the procedure easier. For example:

Note: If your screen doesn't look like the one in Figure 2, hold down the CTRL key and press T twice. Now the layout of the screen should resemble that in the figure.

If your readers may not be familiar with the terminology for a complex device or system, include a glossary of key terms. Preparing a glossary of terms early in the writing also helps you use terms consistently, coordinate a team effort, and ease translation if your manual addresses different language groups.

DESIGNING USABLE INSTRUCTIONS

Because technical writers often think of instructions as workhorse documents, focusing only on their content and often writing them in teams, instructional materials are frequently dull. The text can be dense and uninviting. The voice that comes across the page or screen may be at best deliberately neutral. Sometimes, it scolds or talks down to the reader. Such sober seriousness often fails to engage the reader or build confidence for the task at hand.

Increasingly, U.S. instructional documents attempt to make reading comfortable in the belief that such comfort improves the instruction's effectiveness. Noted earlier, stories and narratives can serve to vivify technical instructions with personality and personal tone.

More recently, technical communicators have closely associated themselves with user experience researchers and specialists who pay attention to both the functionality and the *experience* of the user of a product. This shift from content to contact has created a new focus in the design of instructions so as to ensure both usability and positive user experience. The guidelines shown in the "Designing Instructions with Empathy" box will help you deliver instructions effectively.

DESIGNING INSTRUCTIONS WITH EMPATHY

Usability is a measure of efficiency, effectiveness, and user satisfaction with a product, system, or documentation. To create usable instructions, you need to put the user at the center of your design process, constantly reminding yourself that user experience matters. Although usability cannot be guaranteed, it should be a goal to achieve.

Creating empathetic instructions also means thinking about the potential situation your user might be in when reading your document. Technical content manager Alexandra Gifuni (2019) said that she consults Google Analytics to see when and where users are using instructional materials. Knowing that her users tend to find instructions related to troubleshooting, and that the majority of them are reading on a desktop computer helps her to decide on content (e.g., larger screenshots, longer sentences, etc.).

Visuals

Either with text or on their own, visuals are essential elements in delivering instructions. Visuals without labels adapt themselves well to manuals that serve different language groups. In addition, visuals help show readers something rather than telling them. They thus reduce ambiguity, especially for readers who are more visually than verbally minded. For example, instead of saying "Click on the Open File icon on the toolbar," show the reader

Click

Whenever using visuals on digital or print instructions, be sure to include alt-texts (alternative texts or image descriptions) to provide captions or descriptive content so that users with low vision can still access the instructions with assistive technology (such as a screen reader).

Paragraphs

Some readers prefer to see instructions in units of paragraphs. Using paragraphs also helps you tuck away explanations and nuances and emphasize the flow of the process. With paragraphs of explanation, you can develop the context for the process and include brief scenarios or stories about how others have used the process successfully. But paragraphs are harder to skim than lists and pictures and usually take up more space. The following paragraphs from the City of Lubbock (Texas) Water Department website effectively guide citizens in preventing backflow:

Simple Steps to Prevent Backflow

Guard against cross connections. A garden hose is directly connected to the drinking water in a home, and is the most common cross connection. When using a chemical sprayer that connects to a garden hose or filling a swimming pool, a Hose Bib Vacuum Breaker attached to your hose faucet is required. Hose Bib Vacuum Breakers are inexpensive, widely available and easily screw on to your hose faucet.

Make sure a backflow prevention device is installed on your home sprinkler system. Common devices are Double Check Valve Assemblies and Pressure Vacuum Breakers. If you decide to install a lawn sprinkler system on your property, make sure that the water from your sprinkler system cannot contaminate your drinking water. You are required to install an approved backflow prevention device to prevent possible contamination of the drinking water supply.

(City of Lubbock, 2021, n.p.)

Lists

Reading at their leisure, citizens might like to see the water distribution system explained in paragraphs. Such a description is intended for advance reading, not while they're setting up their home water system. When you intend your instructions to be used during the process, think about using a list. Lists emphasize individual tasks better. A checklist, for example:

- Reinforces the serious consequences of not following a safety procedure, as in an aircraft pilot's checklist of actions before takeoff, and helps readers skim a procedure to see the big picture.
- Provides a sense of accomplishment as users check off each step.
- Reminds users of lessons learned in a tutorial.
- Encourages compliance.

As an example, consider this checklist that guides customer service providers in answering customer complaints:

1. Listen to the complaint. Do not interrupt.
2. Summarize the complaint for the customer to assure your understanding of the main point.
3. If appropriate, apologize.
4. Offer a solution or alternatives.
5. Thank the customer for calling attention to the problem.

For more about designing, see Chapters 10 and 11

Number the steps for ease of reference and note in the introduction the total number of steps. Use bullets if the list indicates steps whose sequence is not significant or for a list of items such as equipment, materials, or requirements.

Plain Language, Verbs, and Other Forms

The U.S. Plain Writing Act of 2010 is a bill that seeks to "enhance citizen access to Government information and services by establishing that Government documents issued to the public must be written clearly, and for other purposes" (Gov Info, 2010). Technical communicators across the world have treated this Act as a guideline for producing usable and accessible instructions. Verbs and word choices in technical manuals should be edited for readability and comprehensibility.

For more about plain language, see Chapter 3

Whether you use paragraph or list form, express the action at the heart of the instructions with verbs chosen from the following options:

- *Active imperative:*

  o *Measure* the elapsed time.
  o *Prepare* the solution.

- *Active indicative:*

  o The technician *measures* the elapsed time.
  o The technician *prepares* the solution.

- *Passive*:

  o The elapsed time *is measured*.
  o The solution *is prepared*.

- *Conditional:*

  o The elapsed time *should* (or *shall*) *be* measured.

  o The technician *should* (or *shall*) prepare the solution.

Choose the form that reflects your corporate style guide, personal taste, customer preference, legal mandates, or tradition. The simplest is the active imperative, which is clear, energetic, and direct. The active indicative is appropriate for some standards and other documents that record a procedure, but the need to repeat the noun (or use a pronoun, which may cause problems in avoiding discriminatory language) makes that usage wordy. Passives often fail to name the agent or actor. The conditional may help emphasize a particular step or action, but, overused, it sounds like scolding. Wherever you can, choose the active imperative.

Once you've chosen a verb form, use it consistently. That consistent usage is a form of parallelism. Parallelism helps you reinforce your message and make your instructions comfortable for the reader. If some item you'd like to include resists such consistent expression, that resistance may be a sign that it isn't a step. For example:

1. Clean the beaker.
2. Place 2 mg of the reagent in the beaker.
3. The temperature in the room should be no more than 25°C.

That third item in the above example is a condition, not a step. It can't be expressed as an imperative verb (except, perhaps, in the form: "Make sure the temperature in the room is no more than 25°C").

At a larger level, maintain parallelism among segments of the instructions. For example, if you write one segment in paragraphs, use paragraphing in all similar segments or explain explicitly why you are changing form. Keep highlighting techniques parallel—for example, box all warnings and accompany them with the same icon.

Users often hang on your every word in a procedure. Make sure those words are exactly right: unambiguous, precise, and standard. One often-told anecdote points out the disastrous effects of ambiguity. After a plane crash, inspectors reviewed the maintenance manual and found this sentence: "Unmount the bolt and replace it." Replace could mean either "put in a new bolt" or "put the old bolt back in." Unfortunately, the ground crew put in the old bolt, and the component—and the plane—failed.

In addition to avoiding ambiguity, be precise. Use operational terms to ensure clarity:

Poor: Respond quickly.
Better: Respond within 1 minute.

Poor: Device operates poorly in a cool room.
Better: Device will not operate below 60°F.

Poor: Buy lots of strawberries.
Better: Buy 4 quarts of strawberries.

Use terms, too, that are standard and accepted in the field. Review dictionaries and thesauruses; scout out the terms your competitors use. Within the document, identify components, devices, and systems by the same term and avoid synonyms. A component labeled as a "knob" in one place should not be called a "handle" someplace else. Avoid acronyms; explain them on the first use when they are necessary.

Delivering Your Instructions

You should design pages to accommodate readers who may face your instructions with a knot in their stomachs. Most users want to get the reading over with—and get on with their work. Follow these guidelines to design effective pages.

Separate Units of Text and Visuals Graphically on the Page

Each step, for example, may constitute a clear unit. Sort through the tasks the user needs to perform to achieve such distinct units. Some researchers recommend a two-page spread in which one page provides a graphic depiction of a task and the other page describes that task in words.

Use Headings and Make Them Work

For example, use questions a reader might ask, and then answer them. Distinguish the headings from the text by highlighting or a change in font. Perhaps use a rule to separate the units. Avoid simply listing capabilities from the point of view of the system or product. Instead, indicate actions from the user's point of view:

Capability	Action
Copy utility	Copying files
Calculator	Adding, subtracting, multiplying, dividing
Special characters	Selecting a font

Differentiate Levels of Information

Different readers ask different questions of a document. Some want to read only the minimum information, perhaps just one or two key points. They'll figure out the rest on their own. Some, too, find a diagram enough; words seem confusing. Other readers seek more background, examples, scenarios, maybe a test to see if they are working right, perhaps alternative strategies for doing the same task. You could customize documents for each reader's needs, but that's expensive, although digital delivery helps, as you'll see. Instead, good design can help you accommodate those differences in one document. Represent the levels of information either by sequence through the document or as a hierarchy.

To follow a sequence:

- Place the most important or simplest functions first, then move into less common—or more complex—applications.
- Provide an overview before segments arranged alphabetically.
- Boldface key terms or cross-referenced items to indicate that more extensive discussion is contained elsewhere in the document.

To show different levels within one unit:

- Color-code items in a table of contents and use that color throughout the document for items at that level. The coding, for example, might represent audiences: white pages for all users, green for mechanics interested in the details, blue for systems operators, and so on.
- Indicate levels through typography—for example, all illustrative examples in Times New Roman 10 point, all notes indented five spaces, all background statements in Arial 10 point, all safety warnings in capital letters.
- Use a graphic symbol to represent information at a certain level.

Accommodate Conditions of Use

Determine where the user is likely to read and incorporate design features that help overcome adverse conditions—for example, large print in a manual to be used in poor light or waterproof pages for manuals in a marine environment.

Keep a Family Look

Make sure all manuals in a series look alike and reflect the image of your organization. In particular, documents for foreign markets should not look second rate. If the manual will be used in a remote location, consider developing a quick reference guide for local use.

Digital Delivery

The units of your instructions may need to be adjusted to work on a screen. Companies are increasingly using digital or mobile delivery, especially through websites and in-app support to reduce printing and storage costs and foster easy and speedy updating. Customers also gain the advantages of key-word searching so they can rapidly find what they need. The greater affordance for interactivity in digital instructions means you can create targeted instructions on a screen without taxing the user with full information (see Figure 17.8). Digital delivery also reduces paper use and is, thus, environmentally friendly.

Digital instructions play an important role in fostering globalization. Digital delivery not only eliminates the high cost of shipping but also reduces the difficulties and delays that still accompany the movement of paper across national borders.

As a writer of online documentation, you will face new challenges in overcoming some user resistance to reading online. But you will also help users overcome

17.8
Digital "Walkthroughs"
Use just enough instructions to guide the user through an interface

their fear of opening a hefty print manual. Digitally, you can hide the extent of the procedure, while the user gets just what is needed, when it is needed. You can also embed links to other instructions.

As shown earlier, video instructions can be an effective delivery method for users who prefer a multimodal experience. Create and host videos online so users can access them from anywhere at any time. Keep videos in short segments and show important steps visually (e.g., close-up frames) and verbally (spoken narration). Use the following guidelines to enhance video usability:

- Coordinate demonstrations with text documentation
- Synchronize spoken narration and animation carefully
- Be faithful to the actual user interface
- Use highlighting to guide attention
- Ensure user control
- Keep file sizes small (Plaisant & Shneiderman, 2005).

The test of good instructions, in physical form or online, is that they work. The reader can use them to successfully complete the task at hand. If, at the same time, your instructions can make it as easy as possible for the reader and address the reader in a voice that is comfortable and engaging, so much the better.

MANAGING COLLABORATIVE EFFORTS

Writing the instructions for today's complex equipment, systems, and techniques usually requires the joint expertise of many people in a team. The team may include product designers, engineers, writers, technical illustrators, marketing specialists, potential users of the product, media specialists, and translators. That approach also allows insights from the process of writing instructions to aid in designing the product. In one situation, for example, mechanics working on the wing of a fighter plane often stepped on a hydraulic line running under their preferred work area. Writers preparing instructions to warn them against stepping there talked with engineers who had placed the line and found that a simple rerouting solved everyone's problem (Lippincott, 1994).

Planning
To coordinate the team, whether it is writing instructions for building a fighter plane or maintaining a software system, prepare a plan for the documentation. Such planning helps even when you write as a single author, but it is particularly important when you write on a team. Make explicit the answers to these and other questions you have been learning how to ask in this chapter:

- What is your goal in writing the instructions?
- Who will use your instructions?
- What is their goal in using them?
- What tasks do you need to perform to complete the instructions?
- What tasks do the users need to perform to use the system or device?

- What is the environment in which they will use the instructions?
- When do you need to deliver the instructions to the user?
- What schedule do you need to follow to make that delivery?
- How much money will it cost to prepare the instructions?
- What component documents are needed (for example, tutorials, reference materials, help screens, helpline)?
- What is the format of the pages or screens?

Usability Testing

Your instructional materials are not finished until you have tested their usability and accuracy with actual users. In Chapter 11, you have learned about the importance of paying attention to user experience in the iterative design process. What makes an instruction set effective is its adherence to user expectations. To achieve this, you should collaborate with representatives of your users in the development of your instructional materials. They can supply valuable feedback on your ongoing draft.

You may set up formal tests of the instruction's usability. In those tests, you measure the users' responses to such elements as:

- Warnings and cautions
- Extent of coverage for each step
- Order of the steps
- Level of difficulty in vocabulary and concepts
- Graphics.

You can measure how long it takes for the users to complete their tasks using your instructions (i.e., time on task). This is a standard measure for efficiency. In addition, you can also keep track of the success and failure rates—and their severity—of people using your instructions to accomplish tasks as part of measuring the accuracy of your instructions.

For more about testing and revising, see Chapter 11

Here are five common techniques for such usability testing:

- *Focus group interview.* Select some six to eight potential users who will meet to discuss the instructions at the outline stage or when you have completed a preliminary draft. Discuss format, problems, and solutions.
- *In-depth interview.* Interview potential users to determine what they want to see in the instructions or to discuss their recommendations after reviewing a draft.
- *Questionnaire:* Circulate a draft and a set of questions to potential users.
- *Expert review.* Interview or circulate a questionnaire to people who are knowledgeable about the users. Manufacturers of home health care devices, for example, should interview health care professionals.
- *Operator (user) performance study.* Ask potential users to perform the procedure. Watch them as they work and ask for their comments about any discrepancies between the manual and what actually needs to be done.

Newer methods include eye-tracking and heat maps to identify how users read instructions (e.g., what's the flow of reading, which part of the instructions do they concentrate on most, etc.). Concurrent think-aloud protocol is also popular, asking the user to work through tasks while articulating what they are doing, thinking, and feeling.

Instructing users is a staple technical communication activity. As a technical professional, you may design instructions to support your products or service. In this chapter, you have learned the purposes of instructions and how to create persuasive yet usable instructional materials. Apply strategies in structuring instructions so that they meet the expectations of the users and help them accomplish their goals. Considering the affordances of print and digital delivery, design instructions that achieve positive user experiences.

CHECKLIST: INSTRUCTIONS

1. Think of the reader as a *user* of your instructions: Make them easy to use
2. Think of the reader as a *customer*. Make the instructions persuasive
3. To achieve these purposes, structure information as follows
    Introduction: background, materials, benefits, theory
    Step-by-step description in the easiest steps
    Adequate warnings and cautions
    Maintenance and troubleshooting suggestions

4. Manage a team effort through an effective document plan
    Prepare an audience analysis and purpose statement
    Determine tasks, assignments, schedule, and costs
    Build in tests for usability and accuracy to make sure the instructions
        work

5. Design instructions to *sound* comfortable
    Meet the readers' preference in text and visuals
    Select verbs carefully
    Keep elements parallel
    Use precise and appropriate diction

6. Design the instructions to *look* welcoming
    Divide steps into easy-to-read-and-follow units
    Clearly identify paths and priorities
    Deliver the instructions in the best form, in print or digitally

7. Design the instructions to *work*
    Collaborate with product designers and users
    Test for usability and accuracy

EXERCISES

1. Obtain a copy of a set of instructions that you think are either excellent or very bad. Develop a list of criteria for good instructions based on your reading of this chapter and apply them to your sample. Then, write an evaluation of the instructions in a memo to your instructor. At your instructor's suggestion, rewrite the instructions to improve their utility. Attach a copy of the instructions (or the original) to your memo.

2. For a novice audience (e.g., a first-year engineering major), describe how to operate a device that measures or calibrates something. First, describe the device (remember explaining—see Chapter 8) and general theory of operation. Justify the need to measure whatever the device measures. Then, move to a step-by-step procedure. Include at least one visual in your instructions.

3. Write a memo to a fellow student in your major that explains how to research the leading journals and other literature sources in your college library or on the web.

4. Write a brief set of instructions for a novice audience (e.g., a first-year international student living away from home) that describes how to plan meals and shop for groceries in the local markets. Assume that your audience has been taking out or eating at home until now and knows nothing about shopping in the local markets.

FOR COLLABORATION

In a team of three students, write a document plan that you would follow in creating a set of instructions to accompany some product, system, or process on campus. At this point, you do not need to create the instructions themselves, but create a *plan* for assessing each of the elements you read about in this chapter. Such plans resemble a proposal (see Chapter 13) in that you are detailing a procedure for approval by someone else in an organization.

For example, assume that your plan details how you will create instructions for using a makerspace studio on campus. In your document plan, cover the following topics:

- Goal and scope of the project in general and the instructions in particular
- Project schedule
- Project budget
- Components of the instructions set (e.g., tutorials, reference materials, help documents, tech support)
- Analysis of the users
- Analysis of the environment for use
- Analysis of the user tasks (frequency and sequence)
- Format of the instruction presentation (print or digital).

> Note, too, how you will develop training materials that users can follow on their own at the studio. The tutorials should engage users in doing their work at the studio. To that end, you might include a representative of the users (i.e., a frequent user of the makerspace) as a consultant to your team.

REFERENCES

Apple. (n.d.). Monitor your heart rate with Apple Watch. Retrieved from https://apple.co/2RGkORB

Backinger, C.L., & Kingsley, P.A. (1993). *Write it right: Recommendations for developing user instructions for medical devices in home health care.* HHS Publication FDA 93–4258. Washington, DC: FDA.

Burnett, R.E. (1997). Readers' presumed preferences in Japanese-language and English-language manuals. Presentation at the 61st Annual Convention of the Association for Business Communication, November 8, Chicago.

City of Lubbock. (2021). Water department: Backflow. Retrieved from https://bit.ly/32ztxYe

De Silva, N.H., & Henderson, P. (2007). Narrative-based writing for coherent technical documents. *Proceedings of ACM SIGDOC '07* (pp. 1–8). Retrieved from https://bit.ly/3swHAs5

Fleming, W. (2020). Writing instructions. In *Technical writing for technicians* (n.p.). Retrieved from https://bit.ly/3gyLgXE

Gifuni, A. (2019, July 25). Technically writing with empathy. GSoft Tech. Retrieved from https://bit.ly/3n7fOkw

Gov Info. (2010). Public Law 111–274—Plain Writing Act. Gov Info. Retrieved from https://bit.ly/3tDmbPg

Lippincott, R. (1994, September 16). Validation vs verification. Email to the TECHWR-L list.

McLaren, T.A. (1993, March). Help systems today—New domain for technical communication. *Intercom*, 5–6.

Plaisant, C., & Shneiderman, B. (2005). Show me! Guidelines for producing recorded demonstrations. *2005 IEEE Symposium on Visual Languages and Human-Centric Computing* (pp. 171–178). doi: 10.1109/VLHCC.2005.57

Poe Alexander, K. (2013). The usability of print and online video instructions. *Technical Communication Quarterly, 22*(3), 237–259.

Zhu, C.G. (1995). Statement in the announcement of the FORUM95 video conference, November 13–15, Dortmund, Germany.

18 Correspondence and Online Conversations

The broad term *correspondence* covers many of the ways professionals (and others) communicate to exchange information and maintain goodwill. Traditionally, that meant letters, a form of correspondence still appropriate in many situations, especially in legal circles. In addition, memos have become a common form of correspondence within organizations, now often overtaken by email. You'll probably use all these genres—email, memos, and letters—in your career. In addition, as a student and later in your career, you'll take advantage of online channels that allow you to correspond with others in ways that merge writing and speaking, reading and listening. In this chapter, you will learn strategies for engaging in all these forms of correspondence. These genres differ from the reporting and instructional genres you read about earlier in Part 4 in their emphasis on the *exchange* of information, often person to person.

PURPOSE AND SITUATIONS

As an individual or as a member of an organization, you may find yourself engaged in one or more of four general situations requiring correspondence.

Business or professional correspondence falls into four essential categories:

- Business-to-business (B2B) communication
- Business-to-consumer (B2C) communication
- Consumer-to-consumer (C2C) communication
- Consumer-to-business (C2B) communication.

When choosing the appropriate genre, voice, and design of your communication product, first identify the situation. That understanding will help you determine the specific purpose you need to address.

B2B Communication

As a technical professional, you may be corresponding with technical professionals from other organizations. Business-to-business communication depicts a situation in which one company interacts with another. This situation is typical in wholesale and retail, where one business partners with another business to develop, manufacture, or sell their products. B2B situations usually call for business inquiry letters, memos of understanding, and emails.

DOI: 10.4324/9781003093763-22

B2C Communication

If you work in a consumer-facing unit, such as customer support or social media management, you may be designing correspondence for public audiences. Business-to-consumers (or customers) communication involves situations where one business connects with its customers through public or private channels. B2C messages are usually about promoting a product or service, addressing an issue involving the business, or responding to customer reactions. The genres of correspondence include customer service calls, letters, or emails; social media messages; and press releases or public announcements.

C2C Communication

You may not have control over what your customers say about your company or organization, but you can learn from their correspondence. Customer-to-customer communication usually happens outside the realm of a business's control. C2C instances tend to be mediated by online or digital platforms such as social media and community forums. In these instances, customers or consumers share information about their experiences with certain products and persuade other customers to take certain actions.

C2B Communication

Because professional communication is not a one-way street, good businesses create channels for customers to provide feedback. As a technical professional, you may be in a position to respond to users or consumers who share their insights, reactions, and suggestions. They may, for example, complain, and you answer their complaint. They may compliment your company on your service or product, and you thank them.

Specific Purpose

Within each of these situations, determine the purpose of your correspondence. This section outlines some common ones.

To Announce

Businesses may use official correspondence such as letters and press releases to announce new products, organizational changes, or statements on a certain issue. This kind of correspondence focuses on a central product or issue and aims to invoke desires for the product or adherence to the position taken by the business on an issue.

To Sell or Promote

B2B or B2C sales letters commonly seek to introduce current or new products to a targeted audience. The aim is to get readers to buy. In B2B settings, these actions may involve companies signing on for partnership or investment. In B2C, sales letters aim to appeal to consumers individually (by addressing them by name).

To Inquire

Inquiry letters are written to elicit information from the recipient. The sender wants an answer to a question or a response to a request. Letters of inquiry are used in B2B and C2B contexts where a client or customer seeks responses from a company or service provider.

To Recommend

As a student, you're likely to have requested a letter of recommendation. You ask seasoned professionals who know you to comment on your qualifications for a job, an academic program, or an award. The purpose of the recommendation letter is to augment your profile by revealing positive traits or accomplishments that quantitative measures (such as test scores and grades) can't reveal. In business settings, recommendation letters can be found in B2B, B2C, C2C (e.g., brand advocates or social media "influencers"), and C2B correspondence.

For more about such letters, see Chapter 12

To Complain

Complaint letters are written to express dissatisfaction with a company or provider, and to request for compensation or repayment. This kind of correspondence can be treated as official documents that are recorded for future reference (to be used in legal proceedings, if necessary). The box shows an example of a customer's complaint.

COMPLAINT LETTER

123 Main Street
Town, TX, 77008

March 10, 2021

Mark Smith
Consumer Relations Director
Furniture Gallery and Showroom
555 Broadway
Cityville, KS, 66214

Dear Mr. Smith:

Re: Broken sofa

On February 5, 2021 I bought a sofa, model number 25811, serial number 850599–4204 at the Furniture Gallery and Showroom located at 1834 W. Elm Ave. Town, TX 77001. I paid $650.00 for the sofa on my credit card. Furniture Gallery delivered the sofa to my home on February 12, 2021.

Unfortunately, your product has not performed well because the sofa is defective. I am disappointed because one of the legs broke off on February 28,

---

2021. The sofa is unsteady and rocks while I sit on it, so it is not comfortable or relaxing. I have not used this sofa in a way that would cause any damage. I returned to the store on March 3 and March 5 but the store manager, Aaron, would not speak to me.

To resolve the problem, I would appreciate it if your company would pick up this sofa, for free, and refund the $650 I paid. Enclosed are copies of my records, including my receipt, delivery invoice, and photos of the broken sofa.

I look forward to your reply and a resolution to my problem and will wait until April 5, 2021 before seeking help from a consumer protection agency or Better Business Bureau. Please contact me at the above address or by phone at (281–555–1234).

Sincerely,

[Signature]
Morgan Wilson

Enclosure(s)

Source: Courtesy of USA.gov. Retrieved from www.usa.gov/complaint-letter

EMAIL

Emails are no stranger to the professional world. Even elementary school students are using emails to correspond with their teachers today. The rise of personal computing devices such as laptops and smartphones makes emails a highly accessible mode of communication. In addition, many email service providers such as Gmail, Microsoft Outlook, Yahoo! Mail, and Apple iCloud Mail offer "free" email accounts to anyone with internet access. Keep in mind, however, in signing up for an email

Pros	Cons
• Extremely fast delivery	• Prone to viruses, which may be spread through attachments, spams (unsolicited emails), or phishing attempts
• Accessible anywhere, anytime, with internet	• Prone to identity theft if personal information is used in emails (such as name, bank card information, phone numbers, addresses, etc.)
• Can be sent to one person or a group of recipients	
• Can include attachments such as word documents, images, videos, URLs, and other file formats	

account, users must provide their personal and contact information such as phone numbers, mailing addresses, age, sex, and so on, which are typically used for third-party marketing purposes. In other words, your personal information is your also currency. Use it wisely.

For more about ethical use of information, see Chapter 3

Advantages and Disadvantages

In school and the workplace, emails allow quick, mediated interactions between senders and recipients. The benefits and constraints of email are shown in the table.

Structure

When composing professional emails, pay attention to the following components of an email:

- *Sender (From:)*: Address of the author of the email, either as an individual or organization (for example, "hello@nestle.com"). Keep in mind that emails can be sent by someone on behalf of the author listed (secretary, list manager, etc.).
- *Recipient (To:)*: Include the address(es), or names, of the intended message recipient(s). Multiple emails should be separated by series commas. This is usually automated by the email application.
- *Carbon copy (Cc:)*: Include the address(es) of the secondary recipient(s). This is usually used for documentation or FYI (for your information) purposes. For example, as a user research specialist, you may send usability test findings to the design team and CC the project manager in an email.
- *Blind carbon Copy (Bcc:)*: Include address(es) that you do not want the primary (To:) and secondary (Cc:) recipients to see. When either of these recipients reply to an email that includes BCCs, the BCC contacts will not receive the reply.
- *Date and time*: Standard email clients will automatically time-stamp the email as it is sent (not received). This allows for quick and relatively easy sorting of emails in the future.
- *Subject*: Use a short, descriptive phrase to preview the content of the email. For marketing or business purposes, you may consider using catchphrases or attention-grabbing words such as "Important!", "Expires tomorrow", or "January updates" to motivate recipients to read your email. Most email clients display the subject line in the preview panel of an email dashboard.
- *Message content*: This is where you compose the main texts for an email. Follow the letter structure described next to write your message (i.e., salutation, paragraphs, signature). Use multimedia content sparingly to avoid cognitive overload.
- *Attachment*: Most email clients allow you to include one or multiple attachments of a common file type (.docx, .pdf, .gif, .jpg, .mov, and others) up to a certain size (such as 25MB for Gmail in 2021). If you need to send something greater than the allowed size, you may include a link to a web folder such as Dropbox, Microsoft OneDrive, or Google Drive.

- *Level of importance*: Email clients such as Microsoft Outlook afford you the ability to set and display the importance level of your emails. Visually or textually, you may mark an email message as high, low, or normal level of importance. This feature may be used to help your recipient prioritize their reading of messages and take appropriate actions.
- *Reply or reply all*: When replying to an email, be sure to check who is included in the recipient section (CC). A regular reply usually sends your response to the sender of the email. A reply-all action sends your response to all the recipients included in the original message. In listservs (such as WPA-L, ATTW-L) or mailing groups, a reply message may be automatically sent to all members of the list or group. Replied messages include a modified subject line, with the addition of "RE:" at the front of the original subject.
- *Forward*: You and your message recipients can forward an email to another recipient who was not included in the original email. When forwarding emails, it is helpful to include your comments or thoughts on the message instead of just a direct forward. Be sure to check if there were attachments in the original email before forwarding it so you can choose to include or exclude those attachments for your next recipient(s). Forwarded messages include a modified subject line, with the addition of "FW:" at the front of the original subject.

When using emails, remember that such correspondence can be archived and accessed by a legal party based on your organization's policies. For instance, at most public or state-funded institutions (such as Texas Tech University), employees are encouraged to include a statement similar to the following in their emails to warn recipients about the nature of the correspondence:

> This is a public, university-owned email account and subject to an Open Records request through the Freedom of Information Act at any time. Correspondence of a personal or political nature should be sent to [my.name@gmail.com].

For more about communicating professionally, see Chapter 3

It is best to keep a professional tone in emails. Humor (i.e., jokes) does not usually land well in written forms, so it is best to keep it out of emails. However, on the appropriate occasion, it is acceptable to show warmth or be congratulatory over emails by using positive visuals or even emojis like these: 🎉 (party popper), 👏 (clapping hands), ☺ (smiling face).

MEMOS

The prevalence of email correspondence today has reduced the attention being paid to memos (short for memorandums). However, memos are still employed in business communication for their durability and formatting flexibility. For companies that deal with huge amounts of digital communications, memos provide an analog alternative to emails and text messages.

Memos are brief, usually no more than one page. Unlike letters, which are generally sent between individuals who are not in the same organization, memos are traditionally in-house documents. Among other topics, memos report on research or field investigations, describe policies and procedures, and circulate the minutes of meetings. Memos provided the original structure for emails today; thus, you may find the following items familiar on a memo:

For more about minutes and brief reports, see Chapter 15

To
From
Date
Subject

Because memos circulate within organizations and should be brief, use a less formal style than in a letter. Feel free to use insider's language and abbreviations and to present information more baldly without the elaboration or regard for polite small talk necessary for outsiders. To encourage brevity, some organizations provide half-sheet memo forms, because writers who face a full page may feel compelled to fill it.

The box shows a sample memo sent from a center director to faculty members at a university announcing a grant opportunity. The memo was sent as an email with a copy of a PDF attached to the email.

MEMORANDUM CALLING FOR APPLICATIONS TO AVAILABLE FUNDING

MEMORANDUM

To: TTU Faculty

From: Dr. Michael Borshuk, Director, The Humanities Center at Texas Tech

Date: February 8, 2021

Subject: The Humanities Center and Office of Research and Innovation Call for Working Group Applications for FY2021

The Humanities Center and the Office of Research and Innovation (OR&I) are pleased to offer funding for interdisciplinary Working Groups to support innovative thinking and a trans-disciplinary research culture at Texas Tech University. As we recognize that strong work and effective collaboration often take time and sustained debate, the Humanities Center and OR&I will fund groups to investigate a subject of common concern over three years of activity. Members should comprise a core group of tenured and/or tenure-track faculty, representing multiple disciplines; the broader group may also include post-docs, graduate students, and other qualified researchers (at least two-thirds of the group must be tenured or tenure-track faculty).

Working Groups can request up to **a total of $7,500** to be distributed over the course of three years. Applicants should indicate in a budget how the funds will be spent each year.

Please see the attached pdf for full application details.

The deadline for this competition is Friday, April 3, 2021, at 11:59pm.

For additional questions about this competition, contact Dr. Michael Borshuk (michael.borshuk@ttu.edu) or Dr. Abigail Swingen (abigail.swingen@ttu.edu).

LETTERS

Letters are formal documents, reserved mainly for correspondence between individuals or between an organization and an individual. Letters demonstrate formality, furnish information, encourage action, influence people, create a permanent record, and maintain relationships. As a technical professional, you need to know how to write effectively in this well-respected genre.

Structure

To make your letter look like a letter, include the units conventional in your organization or expected by your reader. Such conventions differ from country to country and, indeed, are changing in the global economy. This section reviews current practice for letters in the United States:

- *Heading*: The heading gives your address and the date. Use a standard letterhead designed for your business or organization.
- *Inside address*: The inside address identifies the name and address of the recipient. If including a postal address, note carefully the order of elements, especially in non-U.S. addresses.
- *Subject*: Less common on letters than memos, a subject line identifies the topic of the letter. If appropriate, refer to the number of an account, the name of a project, or prior correspondence:

  o   Re: Broken sofa ("Re:" here is a shorthand for "regarding" and not to be confused with "reply").

- *Salutation*: The typical salutation in the U.S. is *Dear*, although that may seem old-fashioned, followed by the recipient's name as it appears in the inside address. If you are writing to a role rather than an individual, use an attention line:

  o   Attention: Customer relations.

- *Body*: The body contains the text of the letter's message.
- *Complimentary close*: Most formal U.S. letters include a single line of closure, either "Sincerely" or "Your truly." Professional colleagues may use a variant such as "Best wishes" or "Kind regards"; "Respectfully yours" is appropriate to show deference to the reader, as in submitting a proposal or final report to a client or supervisor.
- *Signature*: The signature block provides the final authority for the letter as a legal document. You sign the letter, and that signature may appear above your

typed name and title. You may also include the company name. Personalize an otherwise formal letter by signing your nickname if your relationship with the reader warrants.

- *Notation*: The space below the typed signature may include such notations as the initials of a typist, the number of enclosures, or the names of people to whom copies were sent. The abbreviation for copies is traditionally "cc." (an abbreviation left over from the days of carbon copies).

- *Subsequent pages*: Try to keep a letter to one page. But, if you need another page, use no letterhead stationery that matches the first page. You may simply number the pages or head the following pages with the recipient's name, the page number, and the date. Avoid having only a line or two, plus the complimentary close and signature, on a continuation page.

Cultural Considerations

The global economy has muted some cultural distinctions in the structure of letters today. A Japanese business owner reading a letter from an American customer, for example, is probably comfortable with an American approach, even if the letter is translated into Japanese. But differences in approach still remain—and should

---

Jason Myers
P.O. Box 2000
Park City, UT 84060

Viz.

VIZ Learning: Next Steps

Dear Jason:

I just wanted to say thank you for checking out the videos I sent over — and to see if you had any questions? I love what the team in Portsmouth said about giving every student "that a-ha moment." For me, it comes down to making learning real & relevant. That means supporting teachers and giving students access to content they can relate to.

Quality curriculum is a big part of that. But hands-on experience is just as important. Ask yourself:
- Can students visualize themselves doing this work?
- Do they have access to real-world role models to show them what's possible?
- Is there a community pathway for them to succeed beyond school?

This is why we built VIZ Learning. Because we wanted students in every school to visualize their own success, and give them the tools and experiences to achieve it.

I'd love to catch up and share some more examples with you. If you're facing equity challenges in your district and considering how STEM can help, drop me a reply.

Yours truly,

Eleanor

Eleanor Smith
Customer Relations Manager
VIZ Learning

18.1
Letter from a
Customer Relations
Manager

be recognized. Researchers have found that American business letters tend to be reader-oriented, French business letters are writer-oriented, and Japanese letters are "nonperson"-oriented, reflecting the *relationship* between people more than the people themselves (Hinds & Jenkins, 1987). More recently, research has shown that Latin American (Guatemalan) business writers tend not to use buffers when presenting bad news compared with the more indirect approach by North American writers (Conaway & Wardrope, 2004). And when translating between languages, studies found that pragmatic and syntactical errors can arouse negative reactions in professionals (Luijkx et al., 2019).

To address these cultural concerns, Bruce Maylath and Kirk St.Amant (2019) noted that technical communicators need to develop knowledge in translation—not just linguistically, but also stylistically in design and conventions—as well as localization. These skills can only be honed through practice and reflection over time. As well, to account for cross-cultural differences and international communication contexts, St.Amant (2018) suggest a comparison exercise to help you identify expectations from different cultural groups; for instance (pp. 231–232):

For more about communicating with diverse audiences, see Chapter 2

- Compare how organizations in the same sector share parallel information with the same cultural groups
- Compare how organizations in different sectors share parallel online information on different topics with the same culture
- Compare how the same companies design websites for sharing parallel information with other cultures.

SOCIAL MEDIA

The increased popularity of social networking sites has led to the emergence of online communities where businesses and organizations can build strong customer relations. In this section, you will find tips on corresponding within social media environments.

User Forums

For more about user instructions, see Chapter 17

User forums are a great tool for gathering user experience, understanding any issues or frustrations, and providing quick support to users. These forums are usually created by a business or organization and moderated by a team of employees. The correspondence that takes place in forums is less formal than emails or letters that someone might send to a designated person at an organization. Figure 18.2 shows a user forum set up by Autodesk, a software company. User forums can also be organized on social discussion platforms such as Reddit. Figure 18.3 shows the IKEA help and discussion forum on Reddit.

As a representative of a business, keep a professional persona when corresponding with users on these platforms. Do not scold, degrade, or make negative statements about them. Always remember that your correspondence is seen as the representation of your company.

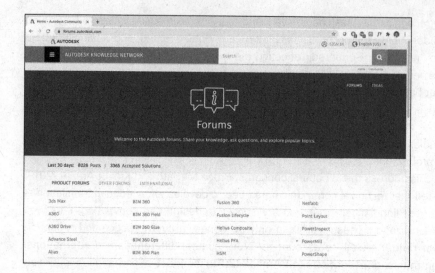

18.2
Autodesk's User
Forum

Source: Retrieved
from https://forums.
autodesk.com

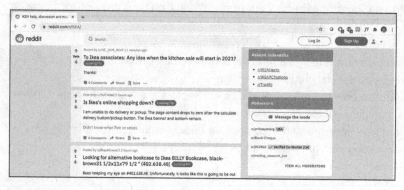

18.3
IKEA's Reddit
Forum

Source: Retrieved from
www.reddit.com/r/IKEA

Official Pages or Accounts

Businesses and organizations may maintain official accounts on social media such as Facebook or Twitter. Like user forums, these platforms allow you to engage with customers virtually.

As shown in Figure 18.4, organizations can promote or sell their products, announce upcoming events, and manage customer complaints—activities that are traditionally done through letters or emails—using social media, for faster and greater reach.

Following these tips, when corresponding with customers over social media:

- *Listen empathetically*: Do not assume customers are intentionally devaluing your business or organization. Show them you care about their experience and are willing to hear their concerns.
- *Respond quickly*: If customers reach out to you on social media, it is because they wish for quick responses. Do not delay your replies.
- *Offer rewards for feedback*: It is kind and beneficial to the image of your business if you extend thank-you messages and small tokens of appreciation for customer feedback. In the Starbucks example (Figure 18.5), a reward of in-store credit was given to a customer who launched a small complaint about the wait time at a store.

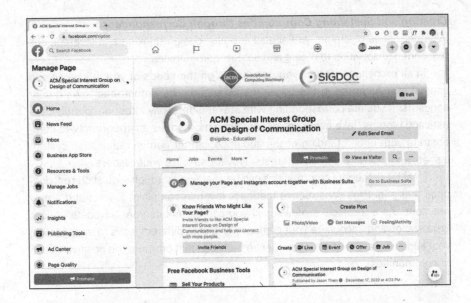

Besides instant messaging, organizations can also install chatbots (or virtual assistants) on social media to automate replies to common questions such as "What time is this store open till?" "Is this product available?" or "Where is the location of this store?" Powered by artificial intelligent applications, chatbots can help companies save time and money by optimizing issue resolution and self-service.

Thanks to machine learning features, which help chatbots to better understand inquiries through big data analysis and adaptations over time, chatbots are becoming more personalized and personable (i.e., corresponding in conversational modes).

Creating and Sustaining Communities through Correspondence

Whether online or offline, you should exercise design thinking and leverage correspondence activities to create and sustain communities.

In all exchanges, center your messages on the needs of your audience and consumers. As noted in the social media section, be empathetic toward user concerns. If you are corresponding with stakeholders, pay attention to their interests and worries. Take every correspondence moment as an opportunity to demonstrate your understanding of the audience, center human rather than system issues, and promote shared interests. Never patronize your readers.

Healthy, respectful communities can be sustained through regular correspondence that prioritizes constructive feedback (even in complaint letters) and timely reactions. As a technical professional, you can be a consumer advocate at your organization and encourage your company to ideate different ways to engage with your audience, and test correspondence methods that promote the positive values of communities.

You probably maintain regular correspondence with your instructors, peers, and colleagues. In this chapter, you are introduced to the situations and purposes for professional correspondences that take in the workplace. To design persuasive and action-driven correspondence, recognize the purpose of your communication and select the appropriate medium. As online correspondence becomes more prevalent, take advantage of social media and community channels to respond to users and customers.

CHECKLIST: CORRESPONDENCE AND ONLINE CONVERSATIONS

1. Before composing any professional correspondence, understand the communication situation
    - B2B communication
    - B2C communication
    - C2C communication
    - C2B communication

2. Determine your purpose for corresponding
    - To announce
    - To sell or promote
    - To inquire
    - To recommend
    - To complain

3. Keep a professional tone when writing an email
4. Use memos to communicate within an organization

5. When composing a letter, do the following
     Identify who you are
     Establish your intentions for writing this letter
     Clarify how the letter fits the context of other correspondence
     Gain the reader's attention by creating common ground
     Promote understanding
     Encourage compliance with your request or decision

6. When responding to a letter
     Answer any direct questions
     Offer a correction or solution, if appropriate
     Be clear but diplomatic in denying a request

7. Use social media sites to build customer relations and maintain communities

---

EXERCISES

1. Write an email requesting information from someone who could provide useful opinions or examples for you to include in a formal report you're working on.
2. As an officer of a student organization, write a letter inviting someone to speak at one of your events or programs.
3. Assuming that the speaker accepted your invitation and participated in your event, prepare a letter to thank them.
4. Compose a memo to members of your student organization about an upcoming special election to fill the position of vice president made vacant by a graduating senior.
5. Write a letter of complaint about a product or service that you found defective. Determine the appropriate person to receive the letter and address that person directly.

FOR COLLABORATION

You and members of your newly formed student organization, TECH Communicators, have been granted permission to create a LinkedIn and a Twitter account. Work with your organization officers to develop a statement of purpose for these social media accounts and create a set of best practices using these platforms. Consider the following questions when working on your statement:

- Who are your primary and secondary audiences?
- What do you aim to do via the social media channels?
- Who will manage these accounts?
- How will interactions with public users be moderated?

- What kinds of content will you publish on these channels? How frequently will you publish new content?
- How will you manage complaints or negative feedback from these channels?
- What will you do to promote a positive outlook of your organization online?

When finished, compose a memo summarizing your efforts and submit it along with your statement of purpose and best practices to your faculty advisor.

REFERENCES

Conaway, R.N., & Wardrope, W.J. (2004). Communication in Latin America: An analysis of Guatemalan business letters. *Business and Professional Communication Quarterly, 67*(4), 465–474.

Hinds, J., & Jenkins, S. (1987). Business letter writing: English, French, and Japanese. *TESOL Quarterly, 21*(2), 327–350.

Luijkx, A., Gerritsen, M., & van Mulken, M. (2019). The effect of Dutch student errors in German business letters on German professionals. *Business and Professional Communication Quarterly, 83*(1), 34–56.

Maylath, B., & St.Amant, K. (Eds.) (2019). *Translation and localization: A guide for technical and professional communicators.* New York, NY: Routledge.

St.Amant, K. (2018). Teaching international and intercultural technical communication: A comparative online credibility approach. In T. Bridgeford (ed.), *Teaching professional and technical communication: A practicum in a book* (pp. 222–236). Louisville, CO: University Press of Colorado/Utah State University Press.

19 Presentations, Posters, and Pitches

As an aspiring technical professional or entrepreneur, you are more than what you *write*. Often, you need to *talk* about your ideas and work at occasions such as business conventions, professional conferences, and public-interest meetings. Presentations are also common in the classroom. Throughout your college career, you will practice delivering oral and poster presentations. In the workplace, you may give in-person or virtual presentations about projects.

If you are like the majority of people on earth, you are probably nervous about giving presentations. It is a common fear. Even game-show hosts and Fortune 500 company executives take hours or days to rehearse and practice their delivery. What you need is a compelling narrative, a story to tell—something that frames your ideas as plausible as well as revolutionary. It may not come naturally, but it can be developed.

To be persuasive, all your communications should be aligned with your story, in particular, your delivery to interested audiences or your pitch to potential investors. You'll also want to persuade brand consultancies to take you on. If so, they can open a gateway to a global marketplace. In this chapter, you will learn the basics of building engaging presentations—in person, via posters, and in situations where you may feel like you're in a shark tank (i.e., business pitches).

VARIED PURPOSES FOR ORAL PRESENTATIONS

Presentations serve to inform the audience about particular findings from a study, give instructions, incite interest in a product, or motivate the audience to take certain actions. This section describes these situations as informative, persuasive, or a combination of the two.

Reporting Findings or Giving Instructions

You may be asked to give an oral presentation that provides information in situations such as the following:

- Reviewing your research with colleagues at a convention or conference
- Briefing members of your class on the major findings in your final report
- Reporting the progress of your project to a board of executives
- Explaining a medical procedure to a group of medical residents.

Your main objective is to provide details and make sure your audience understands what you are saying. This kind of presentation usually manifests in oral or poster mode, where you need to combine verbal and visual content to best deliver specific information. You will learn more about these modes in the upcoming sections of the chapter.

Selling an Idea and Its Future

The second kind of presentation is persuasive as well as informative. Technical professionals often find themselves needing to devise innovative approaches to address existing problems. Design thinking provides a methodology for innovation. You apply that methodology to show how viable and desirable your solution is to an audience panel. In the business world, this kind of presentation is also known as a pitch.

For more about design thinking, see Chapter 1

A business pitch serves to highlight a problem area or entrepreneurial opportunity, describe potential approaches to tackle the problem or opportunity, and introduce a specific solution that can resolve the problem and profit the organization. The ABC show *Shark Tank* (or the BBC version of it, *Dragons' Den*) depicts in an entertainment reality TV manner how entrepreneurial pitches could be given and evaluated. In order to appeal to investors and stakeholders, these pitches are engaging and impressive. Later in this chapter, you will learn how to design and deliver successful pitches.

Inspiring Actions

Your communication professor or instructor may have shown you examples of great presentations from the TED Talk series, a global conference turned non-profit foundation for broad education in science, technology, and the humanities since 1984. Indeed, most presentations from TED and TEDx events embody the essence of the third type of presentation, which is inspirational.

In addition to being persuasive and informative, inspirational presentations combine facts, logic, stories, and community values to motivate actions. In technical and professional communication settings, you may find yourself in a position to inspire others through a presentation when:

- You want to propose a new mission statement for your organization.
- You want to encourage your peers to participate in community service projects.
- You simply want to urge your teammates to complete their work on time.

At any rate, inspirational presentations invoke actions to accomplish small or big goals.

PREPARING A PRESENTATION

An oral presentation is usually given in person, in real time. The first step to preparing your presentation is to create an overview for yourself regarding the scope of the presentation (how much can you cover in the allotted time), audience expectations (who's going to be there and what they need to learn), and presentation aids (what technology or materials you would use).

Scope

The scope of your presentation depends on your goals—what do you want your audience to walk away with? Do you aim to inform, to sell, to inspire, or a combination of these? What would be the ideal effect after your presentation? Alternatively, you may use a simple exercise to determine your main points—fill in the blank:

> After hearing this presentation, my audience should _____.

Outlining your presentation is a smart way to visualize the scope of your presentation and how you manage time. The box shows an example outline of a 15-minute presentation at an academic conference.

OUTLINE OF A 15-MINUTE PRESENTATION AT A SCHOLARLY CONFERENCE

2 mins: Introduce myself and the topic.
3 mins: Describe the background of study. Define problems and rationale.
3 mins: Explain research methods and procedure.
5 mins: Reveal results and two major implications of the study.
2 mins: Close the talk with recommendations for the future.

It is best to keep important information to manageable chunks so that audience members can digest and remember them. For instance, in a 30-minute presentation, you may include three to five key takeaways. In a 15-minute presentation, however, you should keep to one or two main points.

Audience

To help determine your point, and the best way to develop it, consider what your audience expects and needs. Like documents that must work for readers, oral presentations must work for listeners. But keep some key differences in mind. Readers have many options in reading: skimming, returning to the beginning if they get lost, looking at all the pictures and then the text, and reading at their own pace. Listeners have none of these options in real-time presentations. They have to depend on you to highlight a main point and impose a structure and pace that work. Unless your presentation is recorded for playback (which we'll discuss later), listeners can't catch it in the middle and then go back to the beginning. The beginning is gone. If their attention drifts, listeners have missed the information you delivered during that time, and you've lost them. To prevent such losses, make sure you understand your audience.

When someone requests that you give a presentation, ask about the audience. Here are some questions to ask:

- Why is the audience attending the presentation?
- How much do they know about the topic?

- How much do they know about the technical or scientific field as a whole (or the nature of your business or services)?
- What do they expect you to tell them?
- How willing are they to pay attention to your presentation?
- What is the length of presentation they are accustomed to?
- How many people will attend the talk?
- How diverse—in age, nationality, technical skills, experience, and the like—is the group?
- Do all members of the audience speak English fluently (if you're presenting in English)?

If you will be speaking in English, and your listeners will be hearing the talk in another language—that is, if your talk will be interpreted for the local audience—then do the following:

- Prepare visuals that can substitute for text
- Write the main points on a handout, preferably in both English and the language of the audience
- Streamline your vocabulary
- Avoid jokes and metaphors outside the experience of the audience
- Talk with the audience, not to the interpreter, a common problem in such settings.

For more about communicating with diverse audiences, see Chapter 2

Incorporate the answers to these questions in the design of your presentation. For example, if the group will be small, plan to talk informally, with lots of interaction. If the audience is likely to be antagonistic, prepare counterarguments carefully to show that you take their objections seriously and to defuse their opposition. If the audience is diverse, aim at a middle level of understanding in your general remarks, but keep such comments brief to allow time for a question-and-answer period that accommodates different interests. Circulate handouts that provide background information as well as more technical details.

Structure

While an outline gives you an overview of your scope and time management, a structure lets you frame your presentation within an engaging storyline, commonly with an introduction, middle, and end.

Introduction

An old piece of advice about presentation runs, "First you tell 'em what you're gonna tell 'em. Then, you tell 'em, and then you tell 'em what you've told 'em." That's a good approach. In the opening segment, draw the audience's attention and engage their curiosity and interest. For highly motivated or technical audiences, get directly to the point, without wasting their time with preliminaries they will find frustrating. Do, however, provide background for a general audience that needs such information to understand the technical details to follow. Some of the interest-getting devices that work in the introductions of popular science articles serve speakers, too, when they address a general or diverse audience. Consider these openers:

- Mention a local reference point—a person, a place, or a favorite saying, for example—familiar to the audience
- Tell a story or anecdote that relates to your main point
- Open with a question you assume to be on the audience's mind
- Frame the purpose of the talk in terms of what the audience will gain from listening: "By the end of this morning's session, you should be able to assess your company's hiring policies from an ethical vantage point"
- Introduce yourself if you have not already been introduced
- Show a slide that previews the structure or agenda of your presentation.

Middle

In the middle, you tell them. In an informative presentation, you describe or narrate, and you provide instances. In a persuasive or inspirational presentation, you use examples, statistics, expert testimony, and the like to help the audience visualize and commit to the position you advocate. As you do so, keep only the necessary details, because too much information can hinder the audience's memory. Unlike a written report, a presentation aims more at impression than depth—your audience can always read your work later.

For more about reports, see Chapters 15 and 16

Ending

In the end, you make sure the audience knows what your presentation adds up to. You don't want to leave them thinking, "So what?" Prepare a solid ending for your talk, one that will prevent you from rambling and from looking beseechingly at an instructor or other external authority to tell you that you are done. Control the ending in your own terms:

- Clarify any action the audience is supposed to take
- Build motivation for such action
- Broaden the focus
- Recap the main theme
- Summarize the main evidence or components.

CHOOSING PRESENTATION TECHNOLOGY

Choosing and using the right technology for your presentation can enhance the impact of your delivery. Microsoft PowerPoint and other similar slideware (Prezi, Google Slides, Apple's Keynote, etc.) come to mind when presenters think about presentation technology. However, statistician and public speaker Edward Tufte (2003) has called out the diminishing analytical quality of presentations caused by mindless use of slideware templates and bullet points. According to Tufte, presentations that use "pluff"—decorative or low-importance stuff such as ClipArt and animations—can deter audience members from the main message of the presentation. It does not mean that PowerPoint presentations are bad; you just need to be intentional about your design choices.

Designing Slides

It's relatively easy to build a presentation with available slideware. Slides materialize the presentation visually. As people tend to interpret things faster through seeing images than hearing words—especially spoken words that need to be decoded in contexts—slides can augment the message of your presentation by giving your audience something immediate in a visual form. To create impactful presentation slides, consider the following strategies suggested by Canadian-Australian designer Jesse Desjardins (2010):

- *Don't put too much information on a slide.* If your audience is reading what you're saying, what's the point of you being there? Also, don't overdo it with senseless data—put them into meaningful perspectives for your audience.
- *Use strong visuals.* You don't have to use a bunch of stock photos, but use something relevant. Design for the person at the back of the room—that is, make sure they can read your slide, too.
- *Make sure your design (layouts, colors, fonts) is of high quality.* If you don't have the money to hire a professional designer, look up some examples online and imitate them.
- *Don't clutter.* Make room for your content to "breathe" and keep things tidy.
- *Don't underestimate the time it takes to prep.* It's all about being organized and planning ahead. Give yourself plenty of time to design your slides carefully; don't just slap them together at the last minute.

Figure 19.1 is an example of a slide set design that applies the good design principles outlined in Chapter 9 and the advice from Desjardins. These slides applied different font sizes to indicate hierarchy, employed complementary font types to create unity, and used figures to supplement the presenter's oral presentation.

Templates

Although Tufte despised templates, there are some well-designed templates out there that are freely available for you to download and adapt for your own presentation needs. Check out the following providers:

- Canva (www.canva.com/presentations/templates/)
- Slide Carnival (www.slidescarnival.com/category/free-templates)
- Slides Go (https://slidesgo.com/).

Handouts

At professional conferences and business meetings, attendees sometimes expect handouts or something they can reference later on. Especially if your talk is controversial, subject to more than ordinary misunderstanding, or in an

19.1
Slide Set with Clean Layout and Visually Engaging Contents

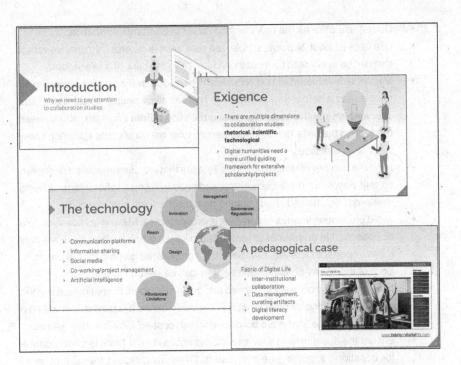

international setting, you may want to provide handouts that allow listeners to follow along while you speak or to work out problems on their own after the talk. Use handouts, then, to:

- Explain a difficult decision or political issue
- Provide a text so the audience is likely to quote you accurately
- Reproduce the main steps in a process for reference when people try the process on their own.

Handouts do not need to always be printed on paper. You may choose to put your presentation notes and summaries on a shareable site such as your own blog or a cloud drive (e.g., Google Drive, Dropbox, etc.) and simply share a link with your audience.

Accessibility

It is important to think about the accessibility of your presentation from the get-go. Accessibility is a huge concern for audience members who have vision or hearing impairments, who find it difficult to follow your presentation if you do not provide accommodations. The following strategies are recommended by Richard Ladner, principal investigator of the AccessComputing Alliance at the University of Washington:

- Minimize the number of slides. No one wants to be shot with a fire hose while trying to understand your talk.
- Use high contrast colors. Audience members with low vision or color blindness will appreciate it.

- Do not use color as the only method for distinguishing information.
- Use large (at least 24 point), simple, san serif fonts (e.g., Arial, Verdana, Helvetica) that can be easily read by most individuals from the back of a large room.
- Minimize the amount of text on slides. When you advance a slide, pause to let people read it before saying anything. This will allow people who are deaf and everyone else in the audience to read the slide before you start talking. Read the text on the slide to make sure people who are blind in the audience know what is on the slide.
- Limit the number of visuals on slides. Images that are used should be described so that people who are blind in the audience will know what image is being displayed. Graphs and charts should be described and summarized.
- Avoid presenting images of complex charts or tables. Make graphics as simple as possible. No one wants to read a complicated graphic when there are only a few important facts about it. Save the complicated graphic for the paper.
- Control the speed of animations so they can be described fully.
- Make sure that videos are captioned and audio described. Sometimes, it is good to give a brief description of what is in the video before it is played. This will help audience members who are blind to establish context for what they will hear.
- Ensure the question and answer period is accessible. If there is a microphone for questioners, make sure they use it. Otherwise, repeat the questions so everyone can hear them.

(DO-IT at University of Washington, 2019, n.p.)

POSTER PRESENTATIONS

Although oral presentations are the most prevalent, another form of presentation that is common at undergraduate, graduate, and professional research conferences is a poster presentation. The poster is a large board (generally around 4' wide by 7' high) that contains your study and most significant findings. In a room filled with long rows of tables, poster boards, or easels, you stand by your poster for a designated period of time. Visitors pass by, browse, and stop to discuss issues with people whose topic interests them.

Abstracts of all the posters and their locations usually appear in a print or digital program for visitors. In preparing a poster, avoid the temptation to merely tape up the pages of a report, but instead design a poster with large print (and high contrast) and visually engaging arrangement to attract the attention of visitors. During the session, because you are probably presenting work in progress, take notes when you hear questions and comments that may help you improve your approach. The Philadelphia Chapter of the Association for Women in Science offers these guidelines for entrants in its competition for student posters on scientific research (courtesy of the Philadelphia Chapter, Association for Women in Science):

- Your audience came to your poster to learn something interesting. Make a good story out of your research. Show the experiments that best allow you to make your point. Sometimes, preparing a presentation helps you see more clearly how to adjust your future research direction.

- Include the following sections: introduction, methods, results, and conclusions. Methods are described in the shortest possible fashion. Supporting and contradictory evidence is part of results, and conclusions should be brief and meaningful.
- Number the poster panels so that anyone can easily follow them in your absence.
- Use colors to make your presentation more exciting.
- Label axes in graphs clearly. Labels in poster presentations should be more extensive than in publications. Multiple traces should be identified directly on the graph rather than in the legend.
- Keep graphs and tables uncluttered. Multiple graphs are preferable to multiple kinds of data on a single graph. Use graphs to show trends in data. Uses tables for precise numerical values. It is easier to compare data in columns than in rows. Have a title above each graph or table.
- Show how reproducible and accurate your results are by including numbers of replicates, standard errors, standard deviations, or confidence intervals in graphs and tables. Include only significant figures.
- Acknowledge the source if you include someone else's data.
- Proofread and check the spelling of your presentation.
- Rehearse aloud and time yourself for an initial 5-minute presentation.
- Allow listeners who are hearing your talk in a non-native language to see as well as hear your words.
- Provide a translation of your main points and the text of your major visuals.

Common software used by poster designers are Adobe InDesign and regular slideware such as Microsoft PowerPoint and Google Slides. In these, you can adjust the size of the poster in the document setup.

A conventional layout for posters is the three-panel design (see Figure 19.2). If you are presenting to a Western audience, the left-to-right, top-to-bottom content flow is suitable for their reading. Figure 19.3 shows an example of such a layout.

19.2
A Three-Panel
Poster Layout

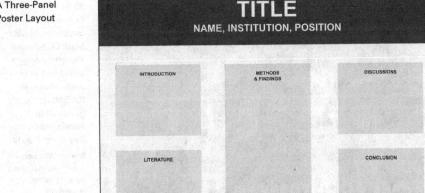

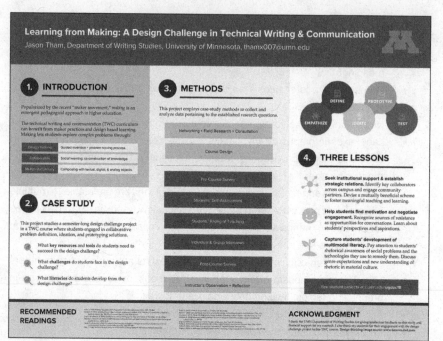

19.3
Poster Design

ENTREPRENEURIAL PITCHES

Pitches are the business version of a proposal presentation. Essentially, it is a chance for you to deliver a well-formed idea, packaged in a persuasive and impressive presentation, to win the buy-in and investment of financiers or venture capitalists so you can launch the idea into the real world. Figure 19.4 shows an example of a pitch in action.

For more about business plans, see Chapter 14

    In order to deliver a successful pitch, you should follow the prior advice given for oral presentations and strategies for designing powerful slides. You should also spend more time on learning about the audience (i.e., investors, funders, venture

19.4
Business Pitch
Sayspring founder Mark Christopher Webster pitching Sayspring in his presentation at the Entrepreneurs Roundtable Accelerator's Demo Day in April 2017

Source: Wikimedia Commons. Retrieved from https://bit.ly/3ehB2Z3

capitalists) who might be interested in backing your project. Spend the time to find out about their investing history, what they find exciting and dull, what kinds of projects are already in their portfolio and in the pipeline.

In an entrepreneurial pitch, you should display enthusiasm and excitement for your idea. You need to show your audience that you, among all people, are the strongest advocate for the idea. But enthusiasm by itself isn't sufficient. Take the following tips curated from the web as well:

- Present your venture idea or new product as though you are presenting it to a 5-year-old. Explain how it works in the simplest terms. Give them the reason they would want to back it, and how it would make the world a better place.
- Have actual representations of potential customers or users. Paint a vivid picture of the demographics of these people for your audience.
- Explain who your competitors are and how you are drastically different from them.
- Explain why you, of all entrepreneurs and existing companies, are the best person to make it happen.
- "Explain what star you can hitch a ride to. Has Best Buy or Radio Shack agreed to distribute your new product? Investors feel much more comfortable knowing you have an established player willing to distribute your wares" (Clary, 2019).
- Be specific about what you're asking and what you're not asking. What's the dollar amount? What are the roles of the investors?
- Explain how the monies will be spent.
- Tell your investors what they can expect in return.

Entrepreneurial or business pitches are different from oral or poster presentations in that you are likely to get questions or requests you may not be ready to answer. As the cliché saying goes, expect the unexpected. You will never be able to anticipate all the things that will happen during a pitch, but you can prepare yourself to receive them. Just know that investors are weary about spending their money on something that isn't already here, so they will try to get as much detail from you as possible. If your audience looks curious and is asking questions, these are good signs that they are interested in your idea.

HARNESS YOUR ENERGY

Always practice your presentation. If possible, practice in the room where you'll be speaking so you can adjust yourself to the equipment available there and become comfortable with the setting. Perhaps assemble a group of friends to simulate an audience and be willing to be part of a practice audience. If your college or other organization offers such a service, videotape your practice session. Reviewing the recording will help you "see yourself as others see you" so you can make appropriate adjustments.

When the day of the presentation arrives, expect to be nervous (even professionals are nervous); then, take a deep breath and forge ahead. Use your nervous energy to deliver what the audience needs. Remember that you are there to achieve something with the audience, at that time and in that space, that is different from writing. Don't waste that occasion. You'll probably be especially nervous during the first few minutes—and especially thankful that you prepared well. As you move through the presentation, let your good preparation help you adjust the talk's structure and pace.

INVOLVE THE AUDIENCE

From the beginning of the talk, develop rapport and common ground with your audience. Talk *with* them, not just *to* them, in a cordial voice that demonstrates your role as part of the group gathered in the room:

- Be flexible. Look for any signs, such as restlessness or puzzled facial expressions, that the audience hasn't understood what you've said. You may need to back up, define a term, or give another example or analogy.
- Slow down. Don't let nervousness make you rush through your presentation or raise the pitch of your voice.
- Avoid slang, jargon not familiar to everyone in the audience, and acronyms. Especially when you address an international audience, use formal English.
- Make sure all pronouns are clear. When you say, "They've developed a promising approach," you may know who "they" are, but your audience may be left wondering.
- Watch for distracting mannerisms. If you've videotaped yourself, you'll see in the playback what these are. Don't push your glasses up on your nose, or tug at a sweater, or move your hands around in your pockets, or push a pointer in and out.
- Maintain eye contact with the audience in cultures such as the United States where that is valued.
- Pause when you need to think and before you say something that is complex or potentially confusing. In general, be confident and show your enthusiasm for the topic. If you don't demonstrate strong interest in what you are saying, it's highly unlikely the audience will be anything but bored.

PRESENTING VIA VIDEO CONFERENCING (SYNCHRONOUS)

At certain occasions, such as a virtual conference or a pitch to international investors, you may need to deliver your presentation using synchronous video conferencing platforms such as Google Meet, WebEx, Skype, or Zoom. Use the following tips when preparing for a virtual presentation:

- Practice using the screen-sharing feature on your videoconferencing application.
- Activate real-time captioning on your slideware to ensure accessibility (Figure 19.5).
- Share your handouts (and slides, if desired) with your audience prior to the presentation.

19.5
Synchronous
Presentation
with Real-Time
Captioning
via Microsoft
PowerPoint

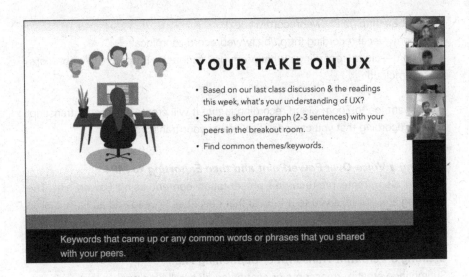

- Slow down your pace. Go a bit slower than you would normally do in face-to-face presentations.
- Pay attention to audience reactions or questions through the chats and other interactive features on your conferencing platform.

RECORDED PRESENTATIONS (ASYNCHRONOUS)

Sometimes, presentations are recorded to be viewed by an audience at a convenient time. When preparing a recorded presentation, follow all the tips for virtual presentations in the previous section and, in addition:

- Make sure your recording is clear (not fuzzy) and playable from any conventional video player or web browser (common file formats include .mp4, .mov, and .wmv).
- Keep the recording short. It is different for viewers to watch a 15-minute video in one sitting than to see a live presentation for the same length of time. The best video presentations are those that are succinct and short.
- Speak clearly and slightly slower than you would in person. Provide an audio transcript for your presentation (see guidelines below).

You can create recorded presentations in the following ways (instructions courtesy of ACM-SIGDOC virtual conference guidelines).

Using a Web Conference Platform to Record Video
One method for recording your presentation is to use a video conferencing technology to record a meeting in which you present with slides. The following links take you to tutorials that describe how to record a meeting using common video conference software.

- Google Meet (http://bit.ly/webrecord-googlemeet)
- Microsoft Teams (http://bit.ly/webrecord-msteams)

- WebEx (http://bit.ly/webrecord-webex)
- Zoom—local recording (http://bit.ly/webrecord-zoomlocal)
- Zoom—cloud recording, with premium accounts (http://bit.ly/webrecord-zoomcloud).

One advantage to Zoom Cloud recording is that it will create a verbatim transcript of your recording that you can adjust to use as your transcript.

Creating a Voice-Over PowerPoint and then Exporting to MP4

You can also create pre-recorded presentation content using PowerPoint. The following link will take you to a tutorial that describes how to add audio to a slide presentation, which can be exported to video format: http://bit.ly/voiceover-msppt

Creating a Verbatim Transcript or Captioning File

Submitting verbatim transcripts or captioning files will help ensure that your content is as accessible as possible to conference attendees. Transcripts are verbatim, word-for-word scripts of the words spoken in your video, while captioning files (such as SRT) include numbers, timestamps, and words spoken. Your video programs may offer integrated support for creating transcripts or captions.

To create a transcript file, type out a word-for-word script of your presentation before recording your video and then adjusting that script after speaking to create your transcript. You can also use a machine-aided transcription service to generate a transcript and adjust the output. Finally, as mentioned previously, Zoom cloud recording will also generate an automatic machine-aided transcript that also can be adjusted as your transcript. Additional resources are available:

- Tips for creating a script before recording: http://bit.ly/screencastomatic-script
- Google Docs: http://bit.ly/googledocs-voicewriting
- Speechnotes: https://speechnotes.co/#
- Zoom: http://bit.ly/zoomcloud-autotranscribe

To create a captioning file in a basic captioning format such as SRT or VTT, you can begin with an existing transcript and add additional content manually in a text editor or by using a free online captioning support program. You can also generate a captioning file from a video that you upload to YouTube:

- Manually creating an SRT file in a text editor: http://bit.ly/createsrtmanually
- Amara: http://bit.ly/amara-captioning
- YouTube: http://bit.ly/youtube-export-autocaption

Audio Poster Walkthrough

If you are presenting a poster asynchronously, consider using an audio recording. To provide context for your poster, you may create an audio file that introduces viewers to your project and to your poster. Many phones include a voice recording app that will enable you to record an audio file directly to your phone (i.e., Voice Memo on iOS; free apps such as Voice Recorder for Android). You may also use Audacity, a free software download that allows you to both record and edit audio. Refer to the Audacity wiki for creating your audio recording: http://bit.ly/audacity-manual

Similar to video recording, you should keep audio recording short and comprehensible with clear voice-overs. Supplement your audio file with a transcript to ensure accessibility.

PRESENTING AS A TEAM

For more about communicating collaboratively, see Chapter 4

When the situation requires a team presentation, preparation is the key. First, agree on the purpose, the genre, and the plan. Make sure everyone knows how long the entire presentation should last and buys into the main point and the best approach. Agree on how you will answer questions. For example, will one person handle all questions, or will you jump in spontaneously? The opening speaker should announce the agreed-on approach, and you should stick to that decision.

Second, assign speakers carefully. Remember that a presentation is a performance, requiring a dynamic interaction with the audience, especially at the beginning and the end. As on a relay team of runners or swimmers, place your strongest presenters first and last. Determine who on the project team should speak. A student team presentation usually requires the participation of everyone. A professional presentation may fall to only one speaker who summarizes the collaboration or may include the expert testimony of people representing different disciplines.

Third, balance a unity of impression against a diversity of voice. The message should be unified around a single point and plan, but the expression of that message should reflect the individual differences among team members. Unlike a collaborative document, which generally requires a consistent style, a presentation leaves room for individuality. In doing so, however, make sure the performance is under control.

Practice where team members will stand when they are not talking, how you will move back and forth to the podium, how you will pass the baton, so to speak, from one speaker to the next. Practice transitions for a smooth handoff and ruthlessly govern the timing of each person's segment. No one should run over. Make sure that presenters who are not speaking show interest in their teammate who is presenting. If you are not interested in each other, you send a signal to the audience that they can be disinterested either.

ANSWERING QUESTIONS

In either a team or individual presentation, if appropriate, encourage questions from the audience. When you deliver prepared remarks, you may offer a question-and-answer session at the end. You or another person who is in charge of the presentation (i.e., a panel moderator) should announce a time limit for questions, just so the session doesn't simply dribble out, and enforce that limit with a remark such as, "We have time for one more question" at the end. Here are some guidelines for answering questions:

- Answer the question asked
- If you don't understand the question, request that it be rephrased or explained
- If you don't know the answer, say that and perhaps suggest where the answer may be found
- If you are addressing a large group, repeat the question (or restate it if the original was roundabout) before you answer so the audience is clear about what you're discussing.

Because most questions occur toward the end of the talk, your answers leave an important impression.

If your presentation was asynchronous, you may ask that your audience send you questions via emails so you can reply to them later.

THE CULTURES OF PRESENTATION

In any form of presentation, you should accommodate the values and expectations shared by people in the culture where you are speaking. That culture may be both corporate and national. Different organizations (and even divisions within organizations) set different norms for presentations. One analyst of those norms, Rae Gorin Cook, labels two major corporate cultures in the United States: "cowboys" and "suits." In cowboy cultures, presenters dress informally and idiosyncratically, encourage interruptions and confrontations, and emphasize spontaneity. You may recognize it also as the Silicon Valley style today. In "suits" settings, by contrast, presenters dress formally (in suits that resemble other presenters' suits), adhere to rigid standards of formal presentation that allow few questions, use visuals extensively and speak from a prepared text, and cultivate an authoritative stance.

Similarly, successful presentations in some companies are low-key, highly technical, and detailed; in others, they are more general, breezier, more likely to dazzle than inform. So, a "one size fits all" approach to presentations won't work. Flexibility in approach is even more necessary in international settings, where values and expected behaviors can vary even more widely. If possible, sit in on several successful presentations before you need to give one and observe closely the elements of proper behavior. Ask someone familiar with the audience and setting about appropriate style.

Cook notes, for example, that Chinese audiences dislike boasting and guessing; they find the speculations that fill many American presentations wasteful. They prefer to speak only when they are positive about what they are saying. Hispanic professionals often welcome direct, emotional confrontation on technical issues—confrontation that may not be appropriate in an American environment where managers prefer to avoid public criticism. How valued is emotional sensitivity in the audience's culture? Should you display your emotions—or hide them? Should you maintain eye contact, as in the United States, or is such contact considered uncomfortable and invasive? In any setting, watch your body language and limit or avoid such sensitive behavior as touching, pointing a finger, or using other hand signs, such as the American "thumbs up" for approval or "OK" sign. People from different cultures may read very different meanings in those signs (Cook, 1996).

In selecting and arranging your information, accommodate your listeners' perspective on what constitutes appropriate content. The American belief in objectivity is not universally shared. Many people draw the line very differently between facts and opinion and between what is private information and what can be said publicly. How open are your listeners to hearing negative information? How direct should your talk be? A good American presentation is usually direct: the main point followed by its support. Some cultures value indirection, with facts and details serving as a ramp into the main point. In some situations, the concept of a "presentation" is itself foreign. Instead, a group asks and answers questions in an associative way to build information as they build personal relationships. Accommodate differences, for example, by allowing people who don't like confrontation or public display to ask questions in private or in writing.

Your work as a technical professional may involve varied situations that require you to deliver presentations. You have learned in this chapter the guidelines for designing presentations and pitches that appeal to audience values. Use a clear structure to make your story or content visible. Choose presentation technology that complements your style and needs. Whenever appropriate, provide handouts as a reference. Most important, exert confidence through audience engagement and proper coordination if you are presenting as a team.

CHECKLIST: PRESENTATIONS, POSTERS, AND PITCHES

1. Determine the purpose and type of presentation needed
    Report findings or give instructions: informative presentation
    Sell an idea and its future: informative and persuasive presentation
    Inspire actions: persuasive presentation

2. Determine the central message and story that reflect the purpose, setting, and audience

3. Analyze the audience
    Needs
    Expectations
    Level of understanding
    Values

4. Structure the talk
    Divide it into units pegged to time
    Create a clear introduction, middle, and, especially, ending

Make the structure obvious

Ruthlessly eliminate details

5. Make the talk *visible*

Select the appropriate technology for in-person, poster, or virtual presentations

Ensure accessibility to your presentation

Use content and structure visuals

Limit the text on a visual

Motivate but don't overload or merely dazzle

6. Use appropriate handouts

Back up controversial information

Provide translation of main points if needed

Give a reference for a complex process

7. Practice your talk

Coordinate the roles, order, and transitions of a team presentation

Make sure the presentation technology works in the setting

Become comfortable with the technology and with speaking out loud

8. Talk with the audience—not at them

EXERCISES

1. Present an 8-minute talk based on a report you prepared for your technical communication class. Don't just read the report out loud. Instead, *adapt* it for an oral presentation in front of a live audience. If your report addressed a specific client or supervisor, then redirect the oral report to the class. You'll need to change the emphasis and content to match the new audience. For example,

    Original report: An international program for improving veterinary practices in Estonia
    Adapted presentation: Commentary on the brief video used in the Estonian training program.

    Original report: The development and design of new signs for the intersection of Interstates 95, 495, and 295
    Adapted presentation: The process for determining the text and location of interstate highway signs.

    Original report: A survey of documentation practices in a Swiss software firm
    Adapted presentation: Differences in teamwork techniques between Swiss and U.S. software engineers.

2. Develop an outline for your presentation. Attach it to a memo addressed to your instructor that covers the following components as you learned about them in this chapter:

- Audience analysis
- Purpose
- Presentation format/modality
- Main point or theme
- Visuals and technology
- Plan of the presentation:

    o Introduction
    o Overview statement
    o Supporting evidence or components
    o Closing remarks

- Anticipated audience questions—and your responses.

FOR COLLABORATION

In a team of two or three, prepare a literature review on some aspects of virtual (synchronous and/or asynchronous) presentations—for example, advice for expert presenters, a review of presentation applications or platforms, advice on the visual design of presentations, accessibility issues, or techniques for structuring virtual presentations. Then, brief the class informally about your findings. You may also elect to create a poster presentation to display your findings.

REFERENCES

Clary, S.D. (2019). 10 tips for an entrepreneurial pitch. Hackernoon. Retrieved from https://bit.ly/3xkXERp

Cook, R.G. (1996). Enhancing the participation of foreign-born professionals in U.S. business and technology. In D.C. Andrews (Ed.), *International dimensions of technical communication* (pp. 5–22). Arlington, TX: The Society for Technical Communication.

Desjardins, J. (2010). You suck at PowerPoint: 5 shocking design mistakes you need to avoid. Slideshare. Retrieved from https://bit.ly/3gsLnE7

DO-IT at University of Washington. (2019). How you can make your presentation accessible. Retrieved from https://bit.ly/3n6rZOz

Tufte, E. (2003). *The cognitive style of PowerPoint: Pitching out corrupts within*. Retrieved from www.edwardtufte.com/tufte/powerpoint

Index

Entries in *italics* refer to figures; entries in **bold** denote tables.